Computer Simulated Experiments for Electronic Devices Using Electronics Workbench Multisim®

Third Edition

Richard H. Berube
Community College of Rhode Island

PEARSON
Prentice Hall

Upper Saddle River, New Jersey
Columbus, Ohio

Editor in Chief: Stephen Helba
Acquisitions Editor: Dennis Williams
Production Editor: Rex Davidson
Design Coordinator: Diane Ernsberger
Cover Designer: Jeff Vanik
Cover art: Digital Vision
Production Manager: Pat Tonneman
Marketing Manager: Ben Leonard

This book was printed and bound by Courier Kendallville, Inc. The cover was printed by Phoenix Color Corp.

Pearson Education Ltd.
Pearson Education Singapore Pte. Ltd.
Pearson Education Canada, Ltd.
Pearson Education—Japan

Pearson Education Australia Pty. Limited
Pearson Education North Asia Ltd.
Pearson Educación de Mexico, S.A. de C.V.
Pearson Education Malaysia Pte. Ltd.

10 9 8 7 6 5 4 3 2 1

ISBN: 0-13-048784-8

Preface

Computer Simulated Experiments for Electronic Devices Using Electronics Workbench Multisim®, Third Edition, is a unique and innovative laboratory manual that uses Multisim to simulate actual laboratory experiments on a computer. Computer simulated experiments do not require extensive laboratory facilities, and a computer provides a safe and cost effective laboratory environment. Circuits can be modified easily with on-screen editing, and analysis results provide faster and better feedback than a series of lab experiments using hardwired circuits.

The experiments are designed to help reinforce the theory learned in an electronic devices course. By answering questions about the results of each experiment, students will develop a clearer understanding of the theory. Also, the interactive nature of these experiments encourages student participation in the learning process, which leads to more effective learning and a longer retention of the theoretical concepts.

A series of troubleshooting problems is included at the end of many of the experiments to help students develop troubleshooting skills. In each troubleshooting problem, the parts bin has been removed to force the student to find a fault or component value by making a series of circuit measurements using only the instruments provided. A solutions manual showing measured data, calculations, answers to the questions, and answers to the troubleshooting problems is available to instructors.

In the third edition, the circuits have been modified and some of the experiments have been changed for the Multisim 2001 circuit simulator. Also in the third edition, the name of the "Preparation" section of each experiment has been changed to "Theory". This "Theory" section has all of the technical information needed to do the calculations and answer the questions in the "Procedure" section without referring to another textbook, and **makes it possible to use this manual as a combination text and lab manual, if desired.** These modifications should make these experiments work better with the new Multisim software.

Richard H. Berube
Email rberube@ccri.edu

Acknowledgments

I wish to thank Professor Vartan Vartanian of the Community College of Rhode Island for his valuable suggestions. I especially appreciate the dedication and talent of the editorial staff of Prentice Hall, particularly Dennis Williams and Rex Davidson for their encouragement to complete this third edition. I am also grateful to Ben Shriver, my copy editor, whose attention to detail was extremely valuable.

I also wish to thank Joe Koenig, President of Electronics Workbench, and his technical staff for developing Electronics Workbench and Multisim and giving me valuable assistance whenever it was needed.

Richard H. Berube

Contents

Introduction

Electronics Workbench Multisim is similar to a workbench in a real laboratory environment, except that **circuits are simulated on a computer** and results are obtained more quickly. All of the components and instruments necessary to create and simulate mixed-mode analog and digital circuits on the computer screen are provided. Using a mouse, you can build a circuit in the central workspace, attach simulated test instruments, simulate actual circuit performance, and display the results on the test instruments. Because **circuit faults can be introduced without destroying or damaging actual components**, more extensive troubleshooting experiments can be performed. Also, faulty components that are deliberately introduced in a circuit simulated on a computer can help make it easier to find the faulty component in an actual circuit. The **Multisim Help menu** has all the information needed to get started using Multisim. Additional notes on using Multisim have been included in the Appendix at the end of this manual.

Each experiment includes a list of **Objectives**, a **Materials** list, a **Theory** section, **circuit diagrams**, a **Procedure** section, and (in almost all) a **Troubleshooting** section. The Procedure section requires you to record measured data, calculate expected values, and answer a series of questions designed to reinforce the theory. The Theory section provides all of the theory and equations needed to complete the procedure. The Troubleshooting section has a series of problems that require using the theory learned in the experiment to find a defective component or determine the value of a component.

The materials list and circuit diagrams make it possible to perform the experiments in a **hardwired laboratory** environment using actual circuit components. If you perform the experiments in a hardwired laboratory, wire the circuit from the circuit diagram and connect the instruments specified. If exact circuit component values are not available, use component values as close as possible to those listed on the parts list. After the circuit is wired and checked, turn on the power, record the data in the space provided, and answer the questions. Due to component tolerances, real laboratory data will not be exactly the same as data obtained on the computer simulation. If a closer correlation between the computer simulation results and the hardwired laboratory results is desired, change the component values in your computer simulation circuit to match the measured component values in your hardwired circuit. The circuit diagrams also make it possible to simulate a hardwired laboratory environment by wiring the circuits on the computer screen using the components from the parts bin.

The **CD-ROM** provided with this manual contains the circuits needed to perform the experiments in this manual, the troubleshooting circuits, and a textbook version of Electronics Workbench Multisim. This CD-ROM is write-protected; therefore, you will not be able to save circuit changes on the disk. (Use **save-as** to save circuit changes to another disk). If you wish to install the full educational version of Electronics Workbench Multisim onto your computer system, it can be obtained from **Electronics Workbench, 111 Peter Street, Suite 801, Toronto, Ontario, Canada M5V2H1. (Tel 1-800-263-5552, or Fax 416-977-1818).**

I

Diodes

Experiments 1–10 involve semiconductor diodes and zener diodes. First, you will analyze a semiconductor diode and a zener diode. Next, you will analyze a half-wave rectifier, a full-wave rectifier, and a bridge rectifier, and determine the effect of adding a filter capacitor. Based on what was learned, you will do some troubleshooting experiments on a voltage-regulated power supply. Finally, you will analyze diode clipper and diode clamper circuits.

The circuits for the experiments in Part I can be found on the enclosed disk in the DIODES subdirectory.

EXPERIMENT

1

Semiconductor Diodes

Name_____

Date_____

Objectives:

1. Compare the diode junction voltage of a forward-biased *pn* junction to the junction voltage of a reverse-biased *pn* junction.
2. Compare the diode current for a forward-biased *pn* junction to the diode current of a reverse-biased *pn* junction.
3. Calculate the forward-biased diode current using the circuit resistance and compare the calculated value with the measured value.
4. Calculate the reverse-biased diode current using the circuit resistance and compare the calculated value with the measured value.
5. Learn how to troubleshoot diode faults by making circuit measurements and drawing conclusions based on the measurement results.

Materials:

One 0–200 mA dc milliammeter
One 0–20 V dc voltmeter
One 10 V dc voltage source
One 1N4001 diode
One 100 Ω resistor

Theory:

When the **anode** of a diode is more positive than the **cathode**, the diode is **forward-biased**. When the cathode is more positive than the anode, the diode is **reverse-biased**. An **ideal diode** acts like a short circuit ($R_F = 0$) when it is forward-biased. A **real diode** has a very low resistance when the forward-bias voltage (V_F) exceeds the **barrier potential** (≈ 0.7 V for a silicon diode). This happens because there are many charge carriers at the *pn* **junction** of a forward-biased diode once the barrier potential is exceeded. This causes the diode to have a very low voltage drop (≈ 0.7 V) when it is forward-biased.

An ideal diode acts like an open circuit ($R_R \rightarrow \infty$) when it is reverse-biased. A real diode has a very high resistance, approximating an open circuit, when it is reverse-biased. This happens because there are very few charge carriers at the *pn* junction of a reverse-biased diode. Because there is very little current flow in a high resistance, there will be very little voltage drop across the resistance external to the diode (R in Figures 1-1 and 1-2). This will cause most of the supply voltage (V_S in Figures 1-1 and 1-2) to be across the diode when the diode is reverse-biased. Because a real diode is not a perfect open circuit when it is reverse-biased, it has a small **leakage current**. A very sensitive microammeter is required to obtain a reading.

When a diode is in series with a resistor, as shown in Figures 1-1 and 1-2, the diode current (I) can be calculated as the voltage (V) across the resistor (R) divided by the resistance (Ohm's law, I = V/R). The voltage across the resistor can be calculated by subtracting the voltage across the diode from the supply voltage.

For Figures 1-1 and 1-2,

$$I_F = \frac{10\,V - V_F}{100\,\Omega}$$

where I_F is the forward-biased diode current and V_F is the forward-biased diode voltage.

$$I_R = \frac{10\,V - V_R}{100\,\Omega}$$

where I_R is the reverse-biased diode current and V_R is the reverse-biased diode voltage.

Figure 1-1 Diode Voltage

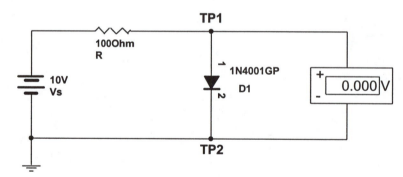

Figure 1-2 Diode Current

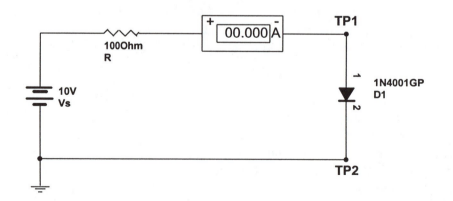

Procedure:

Step 1.   Open circuit file FIG1-1 and run the simulation. The voltmeter is reading the voltage across the diode (diode forward-bias voltage, V_F). Record the reading.

$V_F =$ _____

Question: How does your measured value for the forward-biased diode voltage compare with the typical value for a silicon diode?

Step 2.   Disconnect the diode from the circuit. Select it and rotate it until it is upside-down (reverse-biased). Reconnect the diode to test points TP1 and TP2. Run the simulation again. The voltmeter is reading the diode voltage (reverse-bias voltage, V_R). Record the reading.

$V_R =$ _____

Questions: How does the reverse-biased diode voltage compare with the expected value? **Explain.**

Compare the forward-biased diode voltage and reverse-biased diode voltage. Why are they different?

Step 3.   Calculate the diode current for the forward-biased diode (I_F) from the forward-biased diode voltage and resistor value.

Step 4.   Calculate the diode current for the reverse-biased diode (I_R) from the reverse-biased diode voltage and resistor value.

Step 5.   Open circuit file FIG1-2 and run the simulation. The ammeter is reading the forward-biased diode current (I_F). Record the reading.

$I_F =$ _____

Question: How does your measured value for the forward-biased diode current compare with the calculated value?

Step 6.   Follow the procedure in Step 2 to reverse the diode. Run the simulation again. The ammeter is reading the reverse-biased diode current (I_R). Record the reading.

$I_R =$ _____

Questions: How does your measured value for the reverse-biased diode current compare with the calculated value?

Compare the forward-biased diode current and reverse-biased diode current. Why are they different?

What is reverse-biased leakage current? Is it large enough to be read on your ammeter?

Troubleshooting Problems

1.   Open circuit file FIG1-3 and run the simulation. Based on the voltage across the diode, what is wrong with diode D1?

2.   Open circuit file FIG1-4 and run the simulation. Based on the voltage across the diode, what is wrong with diode D1?

3.   Open circuit file FIG1-5 and run the simulation. Based on the current reading, what is wrong with diode D1?

4.   Open circuit file FIG1-6 and run the simulation. Based on the current reading, what is wrong with diode D1?

EXPERIMENT

2

Semiconductor Diode
Characteristic Curve

Name_____

Date_____

Objectives:

1.   Plot the characteristic curve for the 1N4001 silicon diode.
2.   Demonstrate the forward-bias characteristics of a *pn* junction diode.
3.   Demonstrate the reverse-bias characteristics of a *pn* junction diode.
4.   Compare forward-biased and reverse-biased diode resistance from the characteristic curve.
5.   Compare ac and dc forward-biased diode resistance.
6.   Measure the diode knee voltage (barrier potential) from the characteristic curve.
7.   Plot the diode characteristic curve on the oscilloscope.

Materials:

One function generator
One dual-trace oscilloscope
One 0–20 V dc voltmeter
One 0–100 mA dc ammeter
One 0–5 μA dc microammeter
One 1N4001 diode
One variable 0–20 V dc voltage source
Resistors (one each): 1 Ω, 100 Ω

Theory:

The **diode characteristic curve** can be plotted by measuring the diode voltages for various diode currents and plotting the values on a graph. The test circuit for measuring these values is shown in Figure 2-1. The reversed-biased diode voltages and currents can be measured by reversing the diode in Figure 2-1.

The **ac diode resistance (r)** can be measured by drawing the tangent to the diode characteristic curve and estimating the slope of the tangent line. The ac resistance is equal to the change in diode voltage divided by the change in diode current ($r = \Delta V/\Delta I$), which is equal to the inverse of the slope of the tangent (slope = $\Delta I/\Delta V$). Therefore, the ac diode resistance is equal to the inverse of the slope of the tangent ($r = 1/slope$).

The **dc diode resistance (R)** can be determined directly from Ohm's law by dividing the diode voltage (V) at a particular current by the value of the current ($R = V/I$).

The **diode barrier potential** is the forward-biased diode voltage (V_F) where the diode characteristic curve changes most rapidly (at the knee of the curve).

The diode test circuit for plotting the diode characteristic curve on the oscilloscope is shown in Figure 2-2. In order to plot the diode characteristic curve, the diode voltage must be plotted on the horizontal axis and the diode current must be plotted on the vertical axis. Because the voltage across the 1 Ω resistor (see Figure 2-2) is proportional to the diode current ($I = V/R = V/1$), plotting this voltage on the vertical oscilloscope axis is equivalent to plotting current on the vertical axis. Therefore, because B/A was selected on the oscilloscope, the diode current (I) (terminal B) will be plotted on the vertical axis and diode voltage (V) (terminal A) will be plotted on the horizontal axis. This will produce the I vs. V diode characteristic curve on the oscilloscope screen (Step 13 in the experiment).

Figure 2-1 Diode Test Circuit

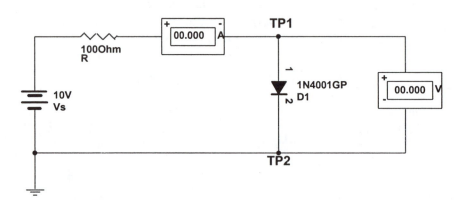

Figure 2-2 Diode Test Circuit for Oscilloscope Curve Plot

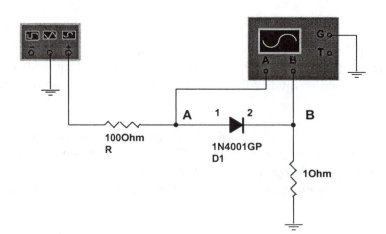

Procedure:

Step 1.   Open circuit file FIG2-1.

Step 2.   Run the simulation. Record the forward-biased diode voltage (V_F) and the forward-biased diode current (I_F) for $V_S = 10$ V in Table 2-1.

Table 2-1

V_S (V)	V_F (V)	I_F (mA)
10		
8		
6		
4		
2		
1		
0		

Step 3.   Change the voltage of the dc voltage source (V_S) to each value in Table 2-1, run the simulation, and record the voltage and current in Table 2-1.

Step 4.   Disconnect the diode from the circuit, select it, and rotate it until it is upside-down (reverse-biased). Reconnect the diode to test points TP1 and TP2. Run the simulation and record the reverse-biased diode voltage (V_R) and reverse-biased diode current (I_R) for $V_S = 0$ V in Table 2-2.

Table 2-2

V_S (V)	V_R (V)	I_R (μA)
0		
5		
10		
15		

Step 5.   Change the source voltage (V_S) to each value in Table 2-2, run the simulation, and record each voltage and current in Table 2-2.

Step 6.   Plot the V_F and I_F data points on the graph for the forward-biased diode. Then draw the curve plot.

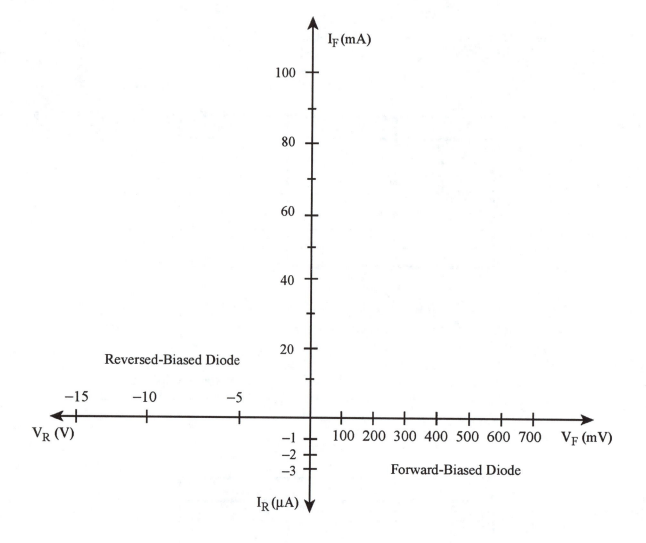

Step 7.   Plot the V_R and I_R data points on the graph for the reverse-biased diode. Then draw the curve plot.

Step 8.   On the forward-biased diode characteristic curve plotted in Step 6, draw the tangent to the curve at $I_F = 20$ mA and estimate the ac forward-biased diode resistance (r_f) from the slope of the tangent. Record the answer.

$r_f = $ _____

Step 9.   From Table 2-2 and the reverse-biased diode characteristic curve plotted in Step 7, estimate the ac reverse-biased diode resistance (r_r). Record your answer.

r_r = _____

Question: Is there a large difference between the ac forward-biased diode resistance (r_f) and the ac reverse-biased diode resistance (r_r)? If so, why?

Step 10.   Calculate the dc forward-biased diode resistance (R_F) at I_F = 20 mA from the equation $R_F = V_F/I_F$.

Question: Is there a difference between the ac and dc forward-bias diode resistance? **Explain.**

Step 11.   Approximate the diode barrier potential from the forward-bias characteristic curve.

Diode barrier potential = _____

Question: Is the diode barrier potential what you would expect for a silicon diode?

Step 12.   Open circuit file FIG2-2. Bring down the oscilloscope enlargement and make sure that the following settings are selected: Time base (B/A), Ch A (Scale = 200 mV/Div, Ypos = 0, DC), Ch B (Scale = 20 mV/Div, Ypos = 0, DC), Trigger (Pos edge, Level = 0 V, Auto). Bring down the function generator enlargement and make sure that the following settings are selected: *Sine Wave*, Freq = 850 Hz, Ampl = 10 V, Offset = 0.

Step 13.   Run the simulation. Notice that you have plotted the diode characteristic curve on the oscilloscope. The voltage across the diode is plotted on the horizontal axis in millivolts (Channel A) and the current through the diode is plotted on the vertical axis in milliamperes (Channel B, 1 mV = 1 mA). Estimate the diode barrier potential from the curve plot.

Diode barrier potential = _____

Questions: How did the shape of the characteristic curve on the oscilloscope screen compare with the shape of the characteristic curve plotted in Steps 6 and 7?

How did the barrier potential in Step 13 compare with the barrier potential in Step 11?

Name_____

Date_____

3

Zener Diodes and Voltage Regulation

Objectives:

1. Plot the reverse-bias characteristic curve for a zener diode and determine the voltage at which breakdown occurs.
2. Demonstrate how the reverse-bias breakdown characteristics of a zener diode can be used to provide a stable dc reference voltage.
3. Calculate zener diode current and zener diode power dissipation.
4. Learn how to determine the ac impedance of a zener diode from the characteristic curve.
5. Demonstrate zener diode voltage regulation with a varying input voltage.
6. Demonstrate zener diode voltage regulation with a varying load resistance.
7. Learn how to troubleshoot zener diode faults by making circuit measurements and drawing conclusions based on the measurement results.

Materials:

One 0–10 V dc voltmeter
One variable dc voltage source
One 1N4733 zener diode
Resistors: 75 Ω, 100 Ω, 125 Ω, 150 Ω, 200 Ω, 300 Ω

Theory:

When a reverse-biased diode reaches **reverse-voltage breakdown,** the reverse-biased diode voltage will remain almost constant with large changes in reverse-bias current. A **zener diode** is a *pn* diode that is designed for operation in the **reverse-bias breakdown region.** If the reverse-bias current (I_Z) goes below a certain level (determined by the knee of the characteristic curve plot), the diode will drop out of the breakdown region and the reverse-bias breakdown voltage (V_Z) will no longer remain constant with variations in zener current (I_Z). The **reverse-bias breakdown voltage (V_Z)** is controlled by the doping level of the *p* and *n* semiconductor. On the zener diode characteristic curve, the reverse-bias breakdown voltage is the value of the voltage where the reverse current begins to increase rapidly (at the knee of the curve plot).

A major application of zener diodes is **voltage regulation.** The circuit in Figure 3-1 will be used to demonstrate zener diode voltage regulation. A zener diode can be used to regulate the output voltage (V_Z) for variations in input (or line) voltage (V_S) or variations in load (R_L). **Percent regulation** is a figure of merit that specifies the performance of a voltage regulator.

The **percent input (or line) regulation** specifies how much the output voltage changes (ΔV_Z) for a given change in the input (or line) voltage (ΔV_S), expressed as a percentage.

$$\text{Percent line regulation} = \frac{\Delta V_Z}{\Delta V_S} \times 100\%$$

The **percent load regulation** specifies how much the output voltage changes (ΔV_Z) for a given change in the load, usually from no load to full load, expressed as a percentage.

$$\text{Percent load regulation} = \frac{(V_{NL} - V_{FL})}{V_{FL}} \times 100\%$$

where V_{NL} is the no load (OPEN) output voltage (V_Z) and V_{FL} is the full load output voltage (V_Z).

In Figure 3-1, the zener diode current (I_Z) can be determined by calculating the current in the 300 Ω resistor following the procedure described in the Theory section of Experiment 1.

$$I_Z = \frac{V_S - V_Z}{300}$$

The zener diode **power dissipation (P_Z)** is calculated by multiplying the zener voltage (V_Z) by the zener current (I_Z).

$$P_Z = V_Z I_Z$$

The ac impedance of the zener diode at a particular point on the diode characteristic curve plot is found by drawing the tangent to the curve and following the procedure in the Theory section of Experiment 2.

Figure 3-1 Zener Diode Test Circuit

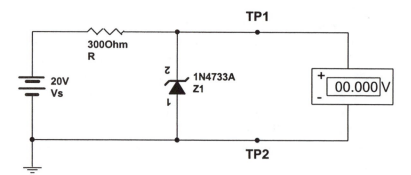

Procedure:

Step 1.   Open circuit file FIG3-1 and run the simulation.   The voltmeter is reading the voltage across the zener diode (V_Z). Record the reading for $V_S = 20$ V in Table 3-1.

Table 3-1

V_S (V)	V_Z (V)	I_Z (mA)
0		
2		
4		
6		
8		
10		
15		
20		
25		

Step 2.   Change the voltage of the dc voltage source (V_S) to each value listed in Table 3-1. Run the simulation, and record the zener voltages (V_Z) in the table for each voltage.

Step 3.   Calculate the zener diode current (I_Z) for each zener voltage value. Record your answers in Table 3-1. Hint: Calculate the current in the 300 Ω resistor.

Step 4.   Plot the V_Z and I_Z data points on the graph and draw the characteristic curve.

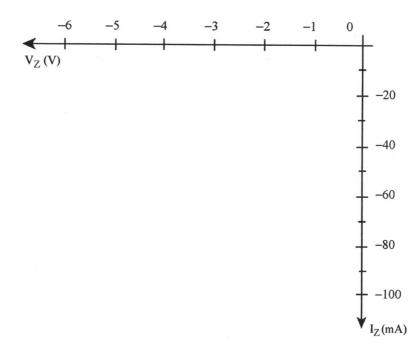

Question: Over what range of values for I_Z was the zener voltage (V_Z) stable?

Step 5.   Estimate the reverse-breakdown voltage for the zener diode from the data on the characteristic curve. Record your answer.

Reverse-breakdown (zener) voltage = _____

Step 6.   Calculate the zener diode power dissipation (P_Z) when the source voltage (V_S) is 20 V.

Step 7.   Based on the slope of the curve plot in the breakdown region, estimate the ac impedance (Z_{ac}) of the zener diode.

Z_{ac} = _____

Step 8.   Connect a 75 Ω resistor (R_L) between test points TP1 and TP2. Change the source voltage (V_S) to 20 V and run the simulation. Record the zener voltage (V_Z) in Table 3-2.

Table 3-2

R_L (Ω)	V_Z (V)	I (mA)	I_L (mA)	I_Z (mA)
75				
100				
125				
150				
200				
300				
OPEN				

Step 9.   Repeat the procedure for each of the resistance values in Table 3-2.

Step 10.   From the zener voltage values determined in Steps 8 and 9, calculate the current in the 300 Ω resistor (I), the load current (I_L), and the zener current (I_Z) for each value of R_L listed in Table 3-2 and record your answers in the table.

Questions: Over what range of values for R_L was the output voltage stable?

What happened to the zener diode current (I_Z) when the load resistance (R_L) went below 125 Ω?

What happened to the zener voltage (V_Z) when the load resistance (R_L) went below 125 Ω? **Explain.**

Step 11.   Based on the data in Table 3-1, calculate the percent input (line) regulation for V_S between 10 V and 20 V.

Step 12.   Based on the data in Table 3-2, calculate the percent load regulation for R_L between OPEN (no load) and 125 Ω (full load).

Troubleshooting Problems

1.   Open circuit file FIG3-2 and run the simulation. Based on the voltage across the diode, what is wrong with diode Z_1?

2.   Open circuit file FIG3-3 and run the simulation. Based on the voltage across the diode, what is wrong with diode Z_1?

3.   Open circuit file FIG3-4 and run the simulation. Based on the output voltage, what is wrong with the circuit? The diode is a 5.1 V zener diode.

EXPERIMENT

4

Half-Wave Rectifier

Name_____

Date_____

Objectives:

1. Demonstrate how a diode can be used to build a half-wave rectifier to convert an ac voltage to a dc voltage.
2. Compare the output voltage waveform with the input voltage waveform for a half-wave rectifier.
3. Calculate the average (dc) value of a half-wave rectified output voltage from the peak output voltage.
4. Determine the dc output ripple frequency for a half-wave rectifier circuit.
5. Demonstrate the effect of diode barrier potential on the half-wave rectified peak output voltage.
6. Determine the peak inverse voltage (PIV) across the diode in a half-wave rectifier circuit.
7. Demonstrate the purpose of the transformer.

Materials:

One 6 V transformer
One 1N4001 silicon diode
One 120 V ac power supply
One 100 Ω resistor
One dual-trace oscilloscope

Theory:

In a **half-wave rectifier**, as shown in Figure 4-1, a load resistance (R_L) is connected to an ac source through a single diode. When the input sine wave is positive, the diode is forward-biased and allows current to flow in the load resistance. When the input sine wave is negative, the diode is reverse-biased and prevents current flow in the load resistance. Therefore, only the positive half of the sine wave voltage cycle is across the load resistance, and the voltage across the load resistance is zero during the negative half of the input sine wave cycle. This produces a pulsating dc voltage across the load resistance. The **average value** of this pulsating half-wave rectified output voltage (V_{dc}) is the value measured by a dc voltmeter, and is calculated by dividing the peak output voltage (V_P) by π. Therefore,

$$V_{dc} = \frac{V_p}{\pi}$$

The diode barrier potential, or forward-bias voltage ($V_F \approx 0.7$ V), causes the peak output voltage to be slightly less than the peak input voltage. This happens because the output voltage is equal to the input voltage minus the voltage drop across the diode.

The maximum value of the reverse-bias voltage across the diode, called the **peak inverse voltage (PIV)**, occurs at the peak of the negative half of the input sine wave. Therefore, the peak inverse voltage (PIV) is equal to the peak value of the input sine wave voltage.

The **output ripple frequency (f)** of this half-wave rectified output is calculated by finding the inverse of the time period (T) for one complete cycle of the output waveshape. Therefore,

$$f = \frac{1}{T}$$

A **transformer** is often used to reduce or increase the ac line voltage to the level desired for the dc output. The transformer secondary peak output voltage (V_{p2}) is equal to the transformer turns ratio times the primary peak input voltage (V_{p1}). Therefore,

$$V_{p2} = \frac{N_2}{N_1} V_{p1}$$

The peak primary (input) voltage (V_{p1}) can be calculated from the rms input voltage using the equation

$$V_{p1} = 1.41 \, (V_{rms})$$

Figure 4-1 Half-Wave Rectifier

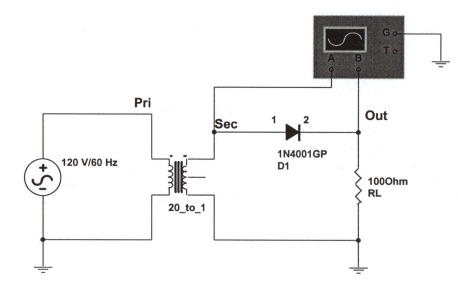

Procedure:

Step 1.   Open circuit file FIG4-1. Bring down the oscilloscope enlargement and make sure that the following settings are selected: Time base (Scale = 5 ms/Div, Xpos = 0, Y/T), Ch A (Scale = 5 V/Div, Ypos = 0, DC), Ch B (Scale = 5 V/Div, Ypos = 0, DC), Trigger (Pos edge, Level = 0 V, Sing, A).

Step 2.   Run the simulation. The blue curve plot on the oscilloscope screen is the output waveform and the red curve plot is the input (Sec) waveform. Draw the input and output curve plots. *Note the peak voltage for each on the graph.*

Questions: What is the main difference between the half-wave rectifier output waveshape and the input waveshape? **Explain why they are different.**

What effect does the diode barrier potential have on the difference between the peak input voltage and the peak output voltage? **Explain why they are different.**

Step 3.   Calculate the average value of the half-wave rectified output voltage (V_{dc}).

Step 4.   Measure the time period (T) for one cycle of the output curve plot and record your answer. From the time period (T), calculate the output ripple frequency (f).

$$T = \underline{\hspace{2cm}}$$

Question: How does the output ripple frequency for this half-wave rectified output compare with the input sine wave frequency?

Step 5.   From the curve plot, determine the largest negative voltage (PIV) across the diode. Record your answer.

Peak Inverse Voltage (PIV) = \underline{\hspace{2cm}}

Question: Does the Peak Inverse Voltage (PIV) across the diode exceed the maximum rating for the 1N4001 diode?

Step 6.   Based on the 120 V rms voltage across the transformer primary, calculate the peak primary voltage (V_{p1}). Using this value and the peak transformer secondary voltage (V_{p2}), calculate the transformer turns ratio.

Question: What is the purpose of the transformer in the half-wave rectifier circuit?

<u>Troubleshooting Problems</u>

1.   Open circuit file FIG4-2 and run the simulation. Based on the curve plot on the oscilloscope screen, what is wrong with the diode?

2.   Open circuit file FIG4-3 and run the simulation. Based on the curve plot on the oscilloscope screen, what is wrong with the diode?

5

Full-Wave Rectifier

Objectives:

1. Demonstrate how diodes can be used to build a full-wave rectifier to convert an ac voltage to a dc voltage.
2. Compare the output voltage waveform with the input voltage waveform for a full-wave rectifier.
3. Compare the output voltage waveform of a full-wave rectifier with the output waveform of a half-wave rectifier.
4. Calculate the average (dc) value of a full-wave rectified output voltage from the peak output voltage.
5. Compare the peak and average (dc) output voltages of a center-tapped full-wave rectifier with the peak and average (dc) output voltages of a half-wave rectifier.
6. Determine the output ripple frequency for a full-wave rectifier circuit and compare it with the output ripple frequency of a half-wave rectifier.
7. Demonstrate the effect of diode barrier potential on the full-wave rectified peak output voltage.
8. Determine the peak inverse voltage (PIV) across each diode in a center-tapped full-wave rectifier circuit.

Materials:

One 6 V center-tapped transformer
Two 1N4001 silicon diodes
One 120 V ac power supply
One 100 Ω resistor
One dual-trace oscilloscope

Theory:

The main difference between a **full-wave rectifier** and a **half-wave rectifier** is that a full-wave rectifier allows current to flow through the load resistor during the entire input sine wave cycle, and a half-wave rectifier allows load resistor current for one half-cycle of the input sine wave. In the **center-tapped full-wave rectifier**, half of the total transformer secondary voltage is between the center tap and each end of the transformer secondary, as shown in Figure 5-1. This causes the full-wave rectifier peak output voltage (V_p) to be approximately one-half the half-wave rectifier peak output voltage, if the transformers have the same turns ratio (20 to 1 in this case). The load resistance (R_L) is connected to the transformer secondary through two diodes. When the secondary sine wave is positive, the top diode is forward-biased and allows current to flow in the load resistance. When the secondary sine wave is negative, the bottom diode is forward-biased and allows current flow in the load resistance in the same direction. This produces a pulsating positive dc voltage across the load resistance for every half cycle of the input sine wave. Because there are twice as many positive pulses in a full-wave rectified output as there are in a half-wave rectified output, the **average voltage (V_{dc})** for a full-wave rectified output is twice that of a half-wave rectified output. Therefore,

$$V_{dc} = \frac{2V_p}{\pi}$$

Because the transformer center tap causes the full-wave rectifier peak output voltage (V_p) to be approximately one-half the half-wave rectifier peak output voltage, the actual average value (V_{dc}) of the rectified full-wave output voltage is approximately the same as the half-wave rectifier average voltage for the same transformer turns ratio.

The diode barrier potential, or **forward-bias voltage (V_F)**, causes the peak of the rectifier output to be less than the peak of the input sine wave (V_{psec1}), because the rectifier output voltage is equal to the input voltage minus the voltage across the diode (≈ 0.7 V).

Figure 5-1 Full-Wave Rectifier

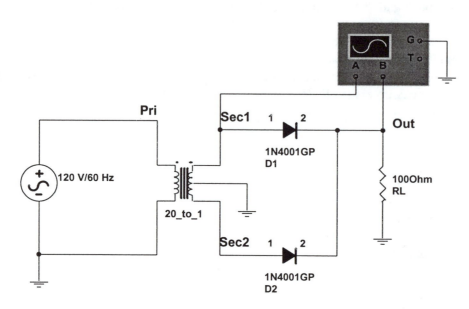

The maximum value of the reverse-bias voltage across each diode, called the **peak inverse voltage (PIV)**, for a center-tapped full-wave rectifier is equal to the total transformer secondary voltage ($V_{psec1} + V_{psec2} = 2\,V_{psec1}$) minus the diode voltage of the forward-biased diode (V_F). Therefore,

$$PIV = 2\,V_{psec1} - V_F$$

The **output ripple frequency (f)** of this full-wave rectified output is calculated by finding the inverse of the time period (T) for one complete cycle of the output waveshape. Therefore,

$$f = \frac{1}{T}$$

Procedure:

Step 1.   Open circuit file FIG5-1. Bring down the oscilloscope enlargement and make sure that the following settings are selected: Time base (Scale = 5 ms/Div, Xpos = 0, Y/T), Ch A (Scale = 2 V/Div, Ypos = 0, DC), Ch B (Scale = 2 V/Div, Ypos = 0, DC), Trigger (Pos edge, Level = 0 V, Sing, A).

Step 2.   Run the simulation. The blue curve plot on the oscilloscope screen is the output waveform and the red curve plot is the input (Sec1) waveform. Draw the input and output curve plots. *Note the peak voltage for each on the graph.*

Questions: What is the main difference between the full-wave rectifier output waveshape and the input waveshape? **Explain why they are different.**

How did the full-wave rectifier output waveshape compare with the half-wave rectifier output waveshape?

How did the peak output voltage of the full-wave center-tapped rectifier compare with the peak output voltage of the half-wave rectifier?

What effect does the diode barrier potential have on the difference between the peak input voltage and the peak output voltage? **Explain.**

Step 3.   Calculate the average value of the full-wave rectified output voltage (V_{dc}).

Question: How did the average value of the full-wave rectified output voltage compare with the average value of the half-wave rectified output?

Step 4.   Measure the time period (T) for one cycle of the output curve plot and record your answer. From the time period (T), calculate the output ripple frequency (f).

          T = _____

Questions: How did the output ripple frequency for this full-wave rectified output compare with the input sine wave frequency? How did it compare with the half-wave output ripple frequency?

Step 5.   From the curve plot, determine the largest negative voltage (PIV) across each diode. Record
                your answer.

                        Peak Inverse Voltage (PIV) = _____

Question: Does the Peak Inverse Voltage (PIV) across the diodes exceed the maximum rating for the
1N4001?

Troubleshooting Problems

1.   Open circuit file FIG5-2 and run the simulation. Based on the curve plot on the oscilloscope
        screen, which diode is defective and what is wrong with it?

2.   Open circuit file FIG5-3 and run the simulation. Based on the curve plot on the oscilloscope
        screen, which diode is defective and what is wrong with it?

3.   Open circuit file FIG5-4 and run the simulation. Based on the curve plot on the oscilloscope
        screen, which diode is defective and what is wrong with it?

4.   Open circuit file FIG5-5 and run the simulation. Based on the curve plot on the oscilloscope
        screen, which diode is defective and what is wrong with it?

6

Bridge Rectifier

Objectives:

1. Demonstrate how diodes can be used to build a bridge rectifier to convert an ac voltage to a dc voltage.
2. Compare the output voltage waveform with the input voltage waveform for a bridge rectifier.
3. Compare the output voltage waveform of a bridge rectifier with the output waveform of a center-tapped full-wave rectifier.
4. Calculate the average (dc) value of the bridge rectifier output voltage from the peak output voltage.
5. Compare the peak and average (dc) output voltages of a bridge rectifier with the peak and average (dc) output voltages of a center-tapped full-wave rectifier.
6. Determine the output ripple frequency for a bridge rectifier circuit and compare it with the output ripple frequency of a center-tapped full-wave rectifier.
7. Determine the peak inverse voltage (PIV) across each diode in a bridge rectifier circuit.

Materials:

One 6 V transformer
Four 1N4001 silicon diodes
One 120 V ac power supply
One 100 Ω resistor
One dual-trace oscilloscope

Theory:

In a **full-wave bridge rectifier**, as shown in Figure 6-1, the load resistance (R_L) is connected to the ac source through four diodes connected as a bridge circuit. When the input sine wave is positive (V_{sec1} higher than V_{sec2}), diodes D_2 and D_3 are forward-biased, allowing current to flow in the load resistance (R_L). When the input sine wave is negative (V_{sec2} higher than V_{sec1}), diodes D1 and D4 are forward-biased, allowing current to flow in the load resistance in the same direction. This produces a pulsating positive dc voltage across the load resistance for every half cycle of the input sine wave. Because there are twice as many positive pulses in this full-wave rectified output as there are in the half-wave rectified output, the **average voltage (V_{dc})** for this full-wave rectified output is twice that of the half-wave rectified output. Therefore,

$$V_{dc} = \frac{2V_P}{\pi}$$

Unlike the center-tapped full-wave rectifier, the bridge rectifier uses the entire transformer secondary. Therefore, the average value of a bridge rectifier output voltage (V_{dc}) is approximately twice as large as the average value of a half-wave rectifier output voltage with the same transformer turns ratio.

As can be seen in Figure 6-1, two diodes are always in series with the load resistance (R_L) during both the positive and the negative half of the input sine wave. Therefore, the **diode barrier potential** (forward-bias voltage) causes the peak of the output to be less than the peak of the input sine wave (V_{sec}). This happens because the output voltage is equal to the input voltage (V_{sec}) minus the voltage drop across the two diodes.

The transformer peak secondary voltage (V_{psec}) is calculated by multiplying the peak primary voltage (V_{ppri}) by the transformer turns ratio. (From Figure 6-1, the ratio of the primary turns to secondary turns is 20:1.) Therefore,

$$V_{psec} = V_{ppri} \frac{N_{sec}}{N_{pri}} = V_{ppri} \frac{1}{20} = \frac{V_{ppri}}{20}$$

Figure 6-1 Bridge Rectifier

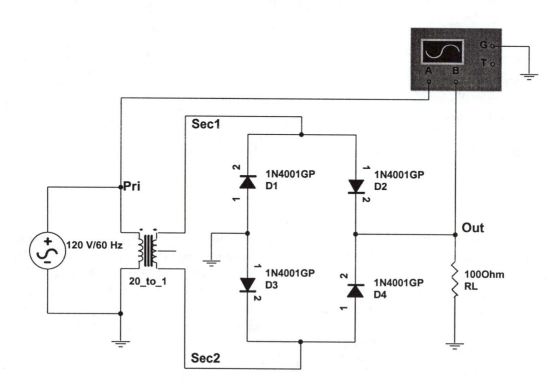

The maximum value of the reverse-bias voltage across each diode, called the **peak inverse voltage (PIV)**, is equal to the transformer peak secondary voltage (V_{psec}) minus the diode voltage of a forward-biased diode (V_F). Therefore,

$$PIV = V_{psec} - V_F$$

The **output ripple frequency (f)** of this full-wave rectified output is calculated by finding the inverse of the time period (T) for one complete cycle of the output waveshape. Therefore,

$$f = \frac{1}{T}$$

Procedure:

Step 1.   Open circuit file FIG6-1. Bring down the oscilloscope enlargement and make sure that the following settings are selected: Time base (Scale = 5 ms/Div, Xpos = 0, Y/T), Ch A (Scale = 100 V/Div, Ypos = 0, DC), Ch B (Scale = 5 V/Div, Ypos = 0, DC), Trigger (Pos edge, Level = 0 V, Sing, A).

Step 2.   Run the simulation. The blue curve plot on the oscilloscope screen is the output waveform and the red curve plot is the input (Pri) waveform. Draw the input and output curve plots. *Note the peak output voltage on the graph.*

Questions: Based on the curve plot, is the bridge rectifier a full-wave or a half-wave rectifier? **Explain why.**

What is the main difference between the bridge rectifier output waveshape and the input waveshape?
Explain.

How does the bridge rectifier output waveshape compare with the center-tapped full-wave rectifier output
waveshape?

How does the peak output voltage of the bridge rectifier compare with the peak output voltage of the
center-tapped full-wave rectifier?

Step 3.   Calculate the average value of the rectified output voltage (V_{dc}).

Question: How does the average value of the bridge rectifier output voltage compare with the average
value of the center-tapped full-wave rectifier output voltage?

Step 4.   Measure the time period (T) for one cycle of the output curve plot and record your answer.
               From the time period (T), calculate the output ripple frequency (f).

               T = _____

Questions: How does the output ripple frequency for this bridge rectifier compare with the input sine wave frequency?   How does it compare with the center-tapped full-wave rectifier output ripple frequency?

Step 5.   From the curve plot, measure the peak primary voltage (V_{ppri}) and record your answer. From this value and the transformer turns ratio, calculate the peak secondary voltage (V_{psec}).

                    Peak primary voltage V_{ppri} = _____

Step 6.   Determine the largest negative voltage (PIV) across each diode. Record your answer.

                    Peak Inverse Voltage (PIV) = _____

Question: Does the Peak Inverse Voltage (PIV) across the diodes exceed the maximum rating for the 1N4001?

Troubleshooting Problems

1.   Open circuit file FIG6-2 and run the simulation. One of the diodes is open. Which diode pair includes the defective diode, based on the curve plot on the oscilloscope screen?

2.   Open circuit file FIG6-3 and run the simulation. One of the diodes is open. Which diode pair includes the defective diode, based on the curve plot on the oscilloscope screen?

Name_____

Date_____

Rectifier Filters

Objectives:

1. Demonstrate the effect on output ripple of a capacitor filter across the output of a half-wave and full-wave rectifier circuit.
2. Measure the peak-to-peak output ripple voltage for a capacitor-filtered rectifier circuit.
3. Calculate the average dc output voltage for a capacitor-filtered rectifier circuit and compare it with the average dc output voltage for an unfiltered rectifier circuit.
4. Compare the value of the average output voltage with the measured dc voltage.
5. Calculate the percent ripple for a capacitor-filtered rectifier circuit.
6. Demonstrate how changing the value of the filter capacitor affects output ripple.
7. Demonstrate how changing the value of the load resistor affects output ripple.
8. Compare the average (dc) output voltage for a capacitor-filtered half-wave rectifier with the average (dc) output voltage for a capacitor-filtered full-wave rectifier.
9. Compare the output ripple voltage for a capacitor-filtered half-wave rectifier with the output ripple voltage for a capacitor-filtered full-wave rectifier.

Materials:

One 6 V center-tapped transformer
Two 1N4001 silicon diodes
One 120 V ac power supply
Capacitors: one 100 μF, one 470 μF
Resistors: one 100 Ω, one 200 Ω
One dual-trace oscilloscope
One multimeter

Theory:

The purpose of a **rectifier filter** is to reduce the ripple in the output voltage to produce a nearly constant dc voltage. A rectifier filter can use capacitance or inductance for filtering, but capacitors are used more often than inductors because large inductors are required and capacitors are less costly. **Capacitor filtering** is accomplished by placing a large capacitor across the rectifier output. This experiment will demonstrate rectifier capacitor filters.

In Figures 7-1 and 7-2, capacitance has been placed across the output of the half-wave and full-wave rectifier circuits from Experiments 4 and 5, respectively. When the rectifier output voltage is rising, the capacitor charges. When the rectifier output voltage tries to decrease between positive peaks, the capacitor discharges slowly. This tends to prevent the output voltage from dropping significantly. If the **RC time constant** of the capacitor (C_1) and the load resistor (R_L) is much greater than the time between positive peaks of the rectifier output, the output voltage will be nearly constant with a small ripple. Increasing the value of the capacitor (C_1) or the load resistor (R_L) causes the RC time constant to increase. This causes the capacitor to discharge more slowly, resulting in less output ripple.

Because the full-wave rectifier **output ripple frequency** is higher than the half-wave rectifier output ripple frequency, a full-wave rectifier is easier to filter and results in a smoother output voltage with less ripple. This happens because the capacitor discharges less during the shorter time interval between positive output voltage peaks in a full-wave rectifier.

Because only half the transformer secondary is used for each half-cycle of the input sine wave in a center-tapped full-wave rectifier, the average output voltage (V_{dc}) will be half the average output voltage of a half-wave rectifier for an equivalent transformer turns ratio. The **average output voltage** (V_{dc}) for a filtered rectifier is calculated by subtracting half the peak-to-peak output ripple voltage (V_{pp}) from the peak output voltage (V_{pout}). Therefore,

$$V_{dc} = V_{pout} - \frac{V_{pp}}{2}$$

The **percent output ripple** is calculated by dividing the peak-to-peak output ripple voltage (V_{pp}) by the average output voltage (V_{dc}) times 100%. Therefore,

$$\% \text{ ripple} = \left(\frac{V_{pp}}{V_{dc}}\right) 100\%$$

Figure 7-1 Half-Wave Rectifier Filter

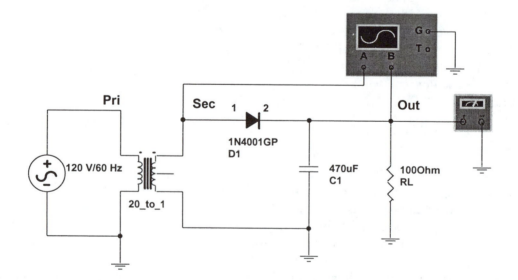

Figure 7-2 Full-Wave Rectifier Filter

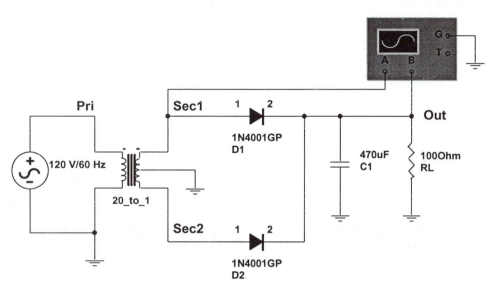

Procedure:

Step 1.   Open circuit file FIG7-1. Bring down the oscilloscope enlargement and make sure that the
             following settings are selected: Time base (Scale = 5 ms/Div, Xpos = 0, Y/T), Ch A (Scale =
             5 V/Div, Ypos = 0, DC), Ch B (Scale = 5 V/Div, Ypos = 0, DC), Trigger (Pos edge, Level =
             0 V, Sing, A). Bring down the multimeter enlargement and make sure that the following
             settings are selected: V, DC (———).

Step 2.   Run the simulation. Pause the simulation after one screen display. The blue curve plot on
             the oscilloscope screen is the output waveform and the red curve plot is the input (Sec)
             waveform. Draw the input and output curve plots. *Note the peak output voltage and the
             peak-to-peak output ripple voltage on the graph.*

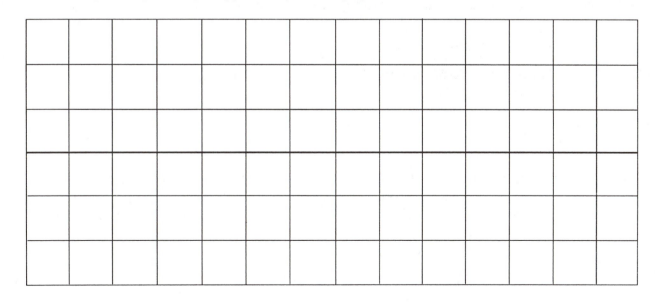

Question: What effect did the filter capacitor have on the output waveshape compared to the output waveshape for the half-wave rectifier without filtering in Experiment 4? **Explain.**

Step 3.   Calculate the average output voltage (V_{dc}) from the peak output voltage (V_{pout}) and the peak-to-peak output ripple voltage (V_{pp}).

Question: What effect did the filter capacitor have on the half-wave rectifier average output voltage compared to the average output voltage without filtering in Experiment 4?

Step 4.   Record the dc output voltage read by the multimeter.

$$V_{dc} = \underline{\hspace{2cm}}$$

Question: How did the measured dc output voltage compare with the calculated average output voltage, V_{dc}?

Step 5.   Calculate the percent ripple from the peak-to-peak output ripple voltage (V_{pp}) and the average output voltage (V_{dc}).

Step 6.   Remove the multimeter. Change the filter capacitor (C_1) value to 100 µF, then run the simulation. Pause the simulation after one screen display. Measure the peak output voltage (V_{pout}) and the peak-to-peak ripple voltage (V_{pp}) on the curve plot and record your answers.

Peak output voltage , $V_{pout} = \underline{\hspace{2cm}}$

Peak-to-peak output ripple voltage, $V_{pp} = \underline{\hspace{2cm}}$

Step 7.   Calculate the average output voltage (V_{dc}) from the peak output voltage (V_{pout}) and the peak-to-peak output ripple voltage (V_{pp}).

Step 8.   Calculate the percent ripple from the peak-to-peak output ripple voltage (V_{pp}) and the average output voltage (V_{dc}).

Question: How did reducing the value of the filter capacitor affect the average (dc) output voltage and the percent ripple? **Explain.**

Step 9.   Change the value of the filter capacitor back to 470 µF and change the value of the load resistor (R_L) to 200 Ω. Run the simulation, then pause the simulation after one screen display. Measure the peak output voltage (V_{pout}) and the peak-to-peak output ripple voltage (V_{pp}) on the curve plot and record your answers.

Peak output voltage ,V_{pout} = _____

Peak-to-peak output ripple voltage, V_{pp} = _____

Step 10.   Calculate the average output voltage (V_{dc}) from the peak output voltage (V_{pout}) and the peak-to-peak output ripple voltage (V_{pp}).

Step 11.   Calculate the percent ripple from the peak-to-peak output ripple voltage (V_{pp}) and the average output voltage (V_{dc}).

Question: How did increasing the value of the load resistance affect the average (dc) output voltage and the percent ripple?   **Explain.**

Step 12.   Open circuit file FIG7-2. Bring down the oscilloscope enlargement and make sure that the following settings are selected: Time base (Scale = 5 ms/Div, Xpos = 0, Y/T), Ch A (Scale = 2 V/Div, Ypos = 0, DC), Ch B (Scale = 2 V/Div, Ypos = 0, DC), Trigger (Pos edge, Level = 0 V, Sing, A).

Step 13.   Run the simulation. The blue curve plot on the oscilloscope screen is the output waveform and the red curve plot is the input (Sec1) waveform. Draw the input and output curve plots. *Measure and record the peak output voltage and the peak-to-peak output ripple voltage on the graph.*

Question: What was the difference between the output waveshape for the full-wave capacitor-filtered rectifier compared to the output waveshape for the half-wave capacitor-filtered rectifier?

Step 14.   Calculate the average output voltage (V_{dc}) from the peak output voltage (V_{pout}) and the peak-to-peak output ripple voltage (V_{pp}).

Question: How did the average (dc) output voltage for the capacitor-filtered center-tapped full-wave rectifier compare with the average (dc) output voltage for the capacitor-filtered half-wave rectifier (Step 3)?

Step 15.   Calculate the percent ripple from the peak-to-peak output ripple voltage (V_{pp}) and the average output voltage (V_{dc}).

Question: How did the percent output ripple for the capacitor-filtered full-wave rectifier compare with the percent output ripple for the capacitor-filtered half-wave rectifier (Step 5)?

Troubleshooting Problems

1.   Open circuit file FIG7-3 and run the simulation. Based on the curve plot on the oscilloscope screen, which diode is defective and what is wrong with it?

2.   Open circuit file FIG7-4 and run the simulation. Based on the curve plot on the oscilloscope screen, which diode is defective and what is wrong with it?

3.   Open circuit file FIG7-5 and run the simulation. Based on the curve plot on the oscilloscope screen, which diode is defective and what is wrong with it?

4.   Open circuit file FIG7-6 and run the simulation. Based on the curve plot on the oscilloscope screen, which diode is defective and what is wrong with it?

8

Voltage-Regulated Power Supply

Objectives:

Learn how to troubleshoot a voltage-regulated power supply by detecting various faults using the electronic instruments available. The power supply specifications are given in Figure 8-1.

Troubleshooting Problems

1. Open circuit file FIG8-1 and run the simulation. Make the appropriate circuit measurements to determine if the power supply is functioning properly. If it does not meet the specifications listed in Figure 8-1, make any circuit measurements desired to determine the defective component and to determine the defect. Remove the defective component and replace it with a new component to determine if your answer is correct.

   Defective component: _____   Defect: _____

2. Open circuit file FIG8-2. Follow the procedure in Step 1.

   Defective component: _____   Defect: _____

3. Open circuit file FIG8-3. Follow the procedure in Step 1.

   Defective component: _____   Defect: _____

4. Open circuit file FIG8-4. Follow the procedure in Step 1.

   Defective component: _____   Defect: _____

Figure 8-1 Voltage-Regulated Power Supply

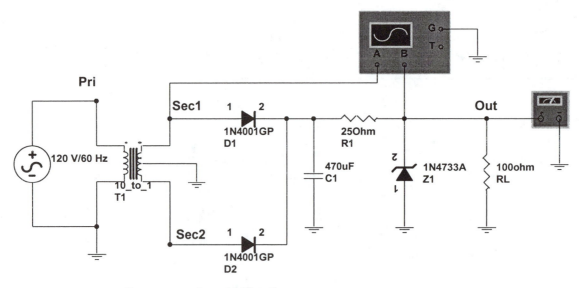

Power supply specifications:
Output voltage = 5.1 V (plus or minus 1%)
Output ripple voltage less than 60 mV p-p
Load resistance cannot be less than 100 Ohm

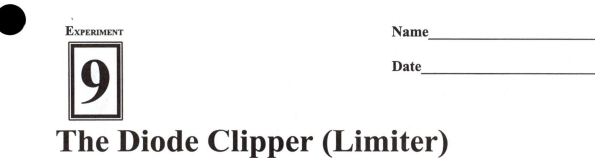

Name_____

Date_____

The Diode Clipper (Limiter)

Objectives:

1. Demonstrate the operation of a series clipper.
2. Demonstrate the operation of a biased series clipper.
3. Demonstrate the operation of a shunt clipper.
4. Demonstrate the operation of a biased shunt clipper.
5. Demonstrate the operation of a zener shunt clipper.
6. Demonstrate the operation of a symmetrical zener shunt clipper.

Materials:

One 5 V dc source
One 1N4001 diode
Two 1N4733 zener diodes
One function generator
One dual-trace oscilloscope
Resistors (one each): 50 Ω, 200 Ω, 500 Ω, 1 kΩ, 5 kΩ, 10 kΩ, 20 kΩ

Theory:

The purpose of a **diode clipper** is to clip off a portion of the input signal voltage above or below a selected voltage level. A diode clipper is sometimes referred to as a **diode limiter**.

The **series clipper**, shown in Figure 9-1, clips off the output at zero volts during the negative part of the input cycle because the diode is cut-off (reverse-biased) during the negative part of the input cycle. The output is not cut-off during the positive part of the input cycle because the diode is forward-biased during the positive part of the input cycle. The peak output voltage is slightly less than the peak input voltage during the positive part of the input cycle because of the diode forward voltage drop (≈ 0.7 V). Reversing the diode would allow only the negative part of the signal voltage to appear at the output.

The **biased series clipper,** shown in Figure 9-2, clips off the output at the bias voltage (5 V) because the diode is cut-off (reverse-biased) when the input voltage is below the bias voltage. The diode conducts only when the input voltage exceeds the bias voltage. Reversing the voltage source (5 V) in Figure 9-2 would make the bias voltage negative.

The **shunt clipper**, shown in Figure 9-3, clips off the output at zero volts minus the diode forward voltage (≈ 0.7 V) during the negative part of the input cycle because the diode is forward-biased during the negative part of the input cycle. The output is not cut-off during the positive part of input cycle because the diode is cut-off (reverse-biased) during the positive part of the input cycle. The peak output voltage is slightly less than the peak input voltage during the positive part of the input cycle because of the voltage division between the 1 kΩ and 5 kΩ resistors. Reversing the diode would allow only the negative part of the signal voltage to appear at the output.

The **biased shunt clipper**, shown in Figure 9-4, clips off the output at the bias voltage (5 V) minus the diode forward voltage (≈ 0.7 V) because the diode is forward-biased when the input voltage is below the bias voltage minus the diode forward voltage (≈ 0.7 V). The diode is cut-off (reverse-biased) only when the input voltage exceeds the bias voltage minus the diode forward voltage (≈ 0.7 V). Reversing the voltage source (5 V) in Figure 9-4 would make the bias voltage negative.

In the **zener shunt clipper**, shown in Figure 9-5, the output voltage cannot exceed the zener reverse breakdown voltage (V_Z) when the input is positive. When the input is negative, the zener diode is forward-biased and the output voltage cannot exceed the forward voltage of the diode (≈ 0.7 V).

In the **symmetrical zener shunt clipper**, shown in Figure 9-6, the output voltage cannot exceed the zener reverse breakdown voltage (V_Z) of the bottom zener diode when the input is positive. When the input is negative, the output voltage cannot exceed the zener reverse breakdown voltage (V_Z) of the top zener diode.

Figure 9-1 Series Clipper

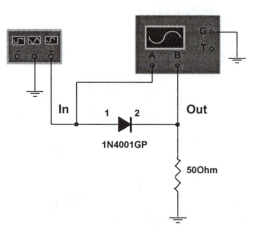

Figure 9-2 Biased Series Clipper

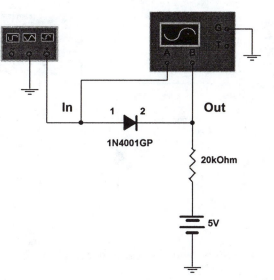

1N4001GP

20kOhm

5V

Figure 9-3 Shunt Clipper

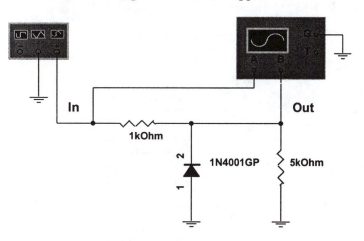

1kOhm

1N4001GP

5kOhm

Figure 9-4 Biased Shunt Clipper

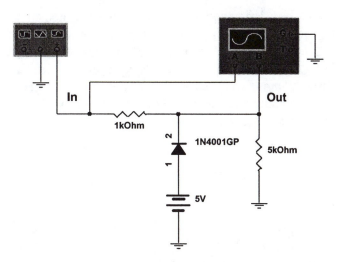

1kOhm

1N4001GP

5kOhm

5V

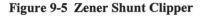

Figure 9-5 Zener Shunt Clipper

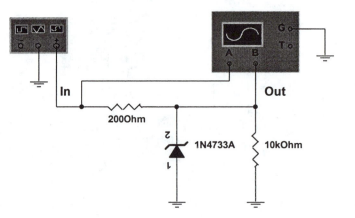

Figure 9-6 Symmetrical Zener Shunt Clipper

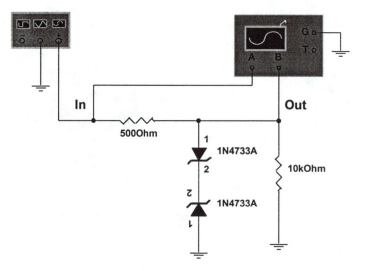

Procedure:

Step 1.   Open circuit file FIG9-1. Bring down the oscilloscope enlargement and make sure that the following settings are selected: Time base (500 µs/Div, Xpos = 0, Y/T), Ch A (Scale = 5 V/Div, Ypos = 0, DC), Ch B (Scale = 5 V/Div, Ypos = 0, DC), Trigger (Pos edge, Level = 0 V, Sing, A). Bring down the function generator enlargement and make sure that the following settings are selected: *Sine Wave*, Freq = 1 kHz, Ampl = 10 V, Offset = 0. Run the simulation. Draw the input and output waveshapes. *Note the peak input voltage, peak output voltage, and clipped voltage level on the graph.*

Question: Why are the peak input and peak output voltages slightly different for the series clipper?

Step 2.   Open circuit file FIG9-2. Make sure that the function generator and oscilloscope settings are the same as in Step 1, then run the simulation. Draw the input and output waveshapes. *Note the clipped voltage level on the graph.*

Question: What is the difference between the clipped voltage level for a series clipper and the biased series clipper? **Explain.**

Step 3.   Disconnect the 5 V source, rotate it until it is reversed, and then reconnect it. Run the simulation. Draw the input and output waveshapes. *Note the clipped voltage level on the graph.*

Question: What determines the clipped voltage level for the biased series clipper?

Step 4.   Open circuit file FIG9-3. Make sure that the function generator and oscilloscope settings are the same as in Step 1, then run the simulation. Draw the input and output waveshapes. *Note the peak input voltage, peak output voltage, and clipped voltage level on the graph.*

<table>
<tr><td></td><td></td><td></td><td></td><td></td><td></td><td></td><td></td><td></td><td></td><td></td><td></td><td></td></tr>
<tr><td></td><td></td><td></td><td></td><td></td><td></td><td></td><td></td><td></td><td></td><td></td><td></td><td></td></tr>
<tr><td></td><td></td><td></td><td></td><td></td><td></td><td></td><td></td><td></td><td></td><td></td><td></td><td></td></tr>
<tr><td></td><td></td><td></td><td></td><td></td><td></td><td></td><td></td><td></td><td></td><td></td><td></td><td></td></tr>
<tr><td></td><td></td><td></td><td></td><td></td><td></td><td></td><td></td><td></td><td></td><td></td><td></td><td></td></tr>
<tr><td></td><td></td><td></td><td></td><td></td><td></td><td></td><td></td><td></td><td></td><td></td><td></td><td></td></tr>
</table>

Questions: What is the difference between the output for the series clipper and the output for the shunt clipper?

What is causing the difference between the peak output voltage and peak input voltage for the shunt clipper?

Step 5.   Open circuit file FIG9-4. Make sure that the function generator and oscilloscope settings are the same as in Step 1, then run the simulation. Draw the input and output waveshapes on the next page. *Note the clipped voltage level on the graph.*

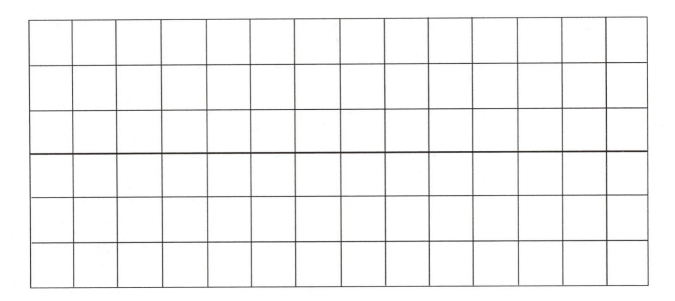

Step 6.   Reverse the 5 V source. Run the simulation. Draw the input and output waveshapes. *Note the clipped voltage level on the graph.*

Question: What determines the clipped voltage level for the biased shunt clipper?

Step 7.   Open circuit file FIG9-5. Make sure that the function generator and oscilloscope settings are the same as in Step 1, then run the simulation. Draw the input and output waveshapes. *Note the peak input voltage, positive clipped voltage level, and negative clipped voltage level on the graph.*

Questions: What determines the positive clipped voltage level for the zener shunt clipper?

What determines the negative clipped voltage level for the zener shunt clipper?

Step 8.   Open circuit file FIG9-6. Make sure that the function generator and oscilloscope settings are the same as in Step 1, then run the simulation. Draw the input and output waveshapes. *Note the positive and negative clipped voltage levels on the graph.*

Question: What is the difference between the zener shunt clipper and the symmetrical zener shunt clipper? **Explain.**

Troubleshooting Problems

1.   Open circuit file FIG9-7 and run the simulation. Based on the curve plot on the oscilloscope, what is wrong with diode D1?

2.   Open circuit file FIG9-8 and run the simulation. Based on the curve plot on the oscilloscope, what is wrong with diode D1?

3.   Open circuit file FIG9-9 and run the simulation. Based on the curve plot on the oscilloscope, what is wrong with diode D1?

4.   Open circuit file FIG9-10 and run the simulation. Based on the curve plot on the oscilloscope, what is wrong with zener diode Z1?

EXPERIMENT

10

Name_____

Date_____

The Diode Clamper

Objectives:

1. Demonstrate the operation of a positive clamper.
2. Demonstrate the operation of a negative clamper.
3. Measure the average value (dc level) of an ac waveform at the output of a diode clamper.
4. Determine the effect of increasing the peak-to-peak ac voltage on the average value (dc level) of the clamper output.
5. Determine the effect of diode forward-bias voltage on the average value (dc level) of the clamper output.

Materials:

One 1N4001 diode
One function generator
One dual-trace oscilloscope
One multimeter
One 10 μF capacitor
One 10 kΩ resistor

Theory:

The purpose of a **diode clamper** circuit is to add a dc voltage level to an ac signal. A **positive diode clamper** adds a positive dc voltage level, and a **negative diode clamper** adds a negative dc voltage level. A clamper circuit is sometimes referred to as a **dc restorer**. Positive and negative diode clamper circuits are shown in Figures 10-1 and 10-2, respectively.

In a diode clamper, when the input voltage increases in one direction, the diode is forward-biased. This allows the capacitor to charge to a voltage near the peak input voltage less the diode forward voltage ($V_{p(in)} - 0.7$ V). When the input voltage increases in the other direction, the diode is reversed-biased. Thereafter, from the peak of one half-cycle to the peak of the other half-cycle the capacitor discharges very little if the RC time constant ($\tau = RC$) is very large compared to the period of the input frequency. The net effect is that the capacitor will retain a charge equal to the peak input voltage less the diode forward voltage, introducing a **dc voltage level** to the ac input.

Figure 10-1 Positive Diode Clamper

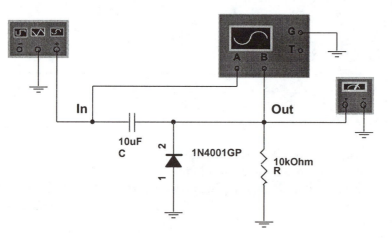

Figure 10-2 Negative Diode Clamper

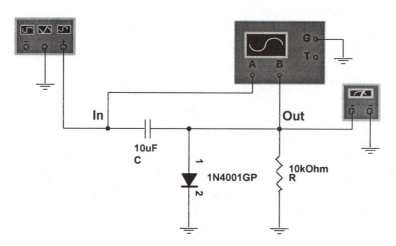

Procedure:

Step 1.   Open circuit file FIG10-1. Bring down the oscilloscope enlargement and make sure that the following settings are selected: Time base (Scale = 500 µs/Div, Xpos = 0, Y/T), Ch A (Scale = 5 V/Div, Ypos = 0, DC), Ch B (Scale = 5 V/Div, Ypos = 0, DC), Trigger (Pos edge, Level = 0 V, Sing, A). Bring down the function generator enlargement and make sure that the following settings are selected: *Sine Wave*, Freq = 1 kHz, Ampl = 5 V, Offset = 0. Bring down the multimeter enlargement and make sure that the following settings are selected: V, DC (———). Run the simulation. After steady state is reached on the multimeter, pause the simulation and draw the input and output waveforms. *Record the peak input and output voltages on the diagram.*

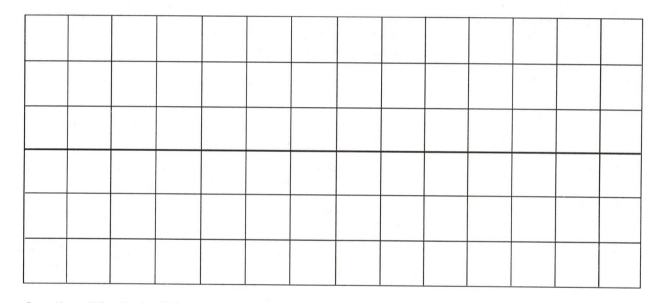

Question: What is the difference between the input and output waveforms for the positive clamper ? **Explain.**

Step 2.   Determine the average value (dc level) of the ac output waveform (blue) on the oscilloscope screen. Record your answer. Also record the multimeter dc voltage reading.

Average (dc level) voltage (oscilloscope): _____

DC voltage (multimeter): _____

Questions: What determines the average value (dc level) of the output waveform for the positive clamper?

How did the average value of the output waveform for the positive clamper compare to the dc voltage reading on the multimeter?

Step 3.   Change the amplitude of the function generator to 8 V, then run the simulation. After steady state is reached on the multimeter, pause the simulation and draw the input and output waveforms. *Record the peak input and output voltages on the diagram.*

Step 4.   Determine the average value (dc level) of the ac output waveform (blue) on the oscilloscope screen. Record your answer. Also record the multimeter dc voltage reading.

                Average (dc level) voltage (oscilloscope): _____

                DC voltage (multimeter): _____

Question: What happened to the average value (dc level) of the output waveform for the positive clamper when the ac input was increased? **Explain.**

Step 5.   Change the amplitude of the function generator to 2 V and the oscilloscope Channel A and
               Channel B to 2 V/Div, then run the simulation. After steady state is reached on the
               multimeter, pause the simulation and draw the input and output waveforms. *Record the peak
               input and output voltages on the diagram.*

Step 6.   Determine the average value (dc level) of the ac output waveform (blue) on the oscilloscope
               screen. Record your answer. Also record the multimeter dc voltage reading.

               Average (dc level) voltage (oscilloscope): _____

               DC voltage (multimeter): _____

Questions: What happened to the output waveform for the positive clamper when the ac input was
decreased to a very low value? Explain the effect of the diode forward voltage on the results.

Why did the diode forward voltage have a greater effect on the positive clamper output for a low ac input
compared to a high ac input?

Step 7.   Open circuit file FIG10-2. Make sure that the function generator, oscilloscope, and multimeter are the same as in Step 1, then run the simulation. After steady state is reached on the multimeter, pause the simulation and draw the input and output waveforms. *Record the peak input and output voltages on the diagram.*

Question: Compare the output of the negative clamper to the output of the positive clamper (Step 1).

Step 8.   Determine the average value (dc level) of the ac output waveform (blue) on the oscilloscope screen. Record your answer. Also record the multimeter dc voltage reading.

Average (dc level) voltage (oscilloscope): _____

DC voltage (multimeter): _____

Questions: Compare the average value (dc level) of the output of the negative clamper to the average value (dc level) of the output of the positive clamper (Step 2).

How did the dc voltage reading on the multimeter compare with the average output voltage on the oscilloscope?

Troubleshooting Problems

1. Open circuit file FIG10-3 and run the simulation. Based on the curve plot on the oscilloscope, what is wrong with the diode?

2. Open circuit file FIG10-4 and run the simulation. Based on the curve plot on the oscilloscope, what is wrong with the diode?

3. Open circuit file FIG10-5 and run the simulation. Based on the curve plot on the oscilloscope, what is wrong with the capacitor?

4. Open circuit file FIG10-6 and run the simulation. Based on the curve plot on the oscilloscope, what is wrong with the capacitor?

II

Bipolar Transistors

The following twelve experiments involve bipolar junction transistors. First, you will plot the characteristic curves for the bipolar junction transistor that will be used in the experiments in Part II. Next, you will study different biasing networks and determine their operating point stability. Finally, you will analyze a small-signal common-emitter amplifier, an emitter-follower amplifier, large-signal class A, class B push-pull, and class C amplifiers, and a cascaded common-emitter amplifier.

The circuits for the experiments in Part II can be found on the enclosed disk in the TRANSIS subdirectory.

EXPERIMENT

11

Name_____

Date_____

Bipolar Transistor Characteristics

Objectives:

1. Demonstrate the relationship between collector current, base current, and collector-emitter voltage for a bipolar junction transistor (BJT).
2. Calculate the dc current gain (β_{DC}) and show how it varies with changes in collector current.
3. Measure the collector leakage current in the transistor cutoff region.
4. Plot the collector characteristic curves for a BJT.
5. Plot the base characteristic curve for a BJT.
6. Plot the emitter current (I_E) versus the base-emitter voltage (V_{BE}) curve for a BJT.
7. Determine the ac current gain (β_{ac}) for a BJT.
8. Determine the ac input resistance from the input characteristics and show how ac input resistance varies with base current.
9. Determine the ac emitter resistance (r_e) from the emitter current versus base-emitter voltage curve and show how ac emitter resistance varies with emitter current.

Materials:

One 2N3904 bipolar junction transistor
Two variable-voltage dc power supplies
One 0–20 mA dc milliammeter
One 0–50 µA dc microammeter
Two 0–25 V dc voltmeters
One 100 kΩ resistor

Theory:

The **bipolar junction transistor (BJT)** consists of three doped semiconductor regions separated by two **pn junctions**. The three regions are called the **collector**, **base**, and **emitter**, with the base being the middle region. There are two types of BJTs. One type consists of two *n* regions separated by a *p* region (*npn* type) and the other consists of two *p* regions separated by an *n* region (*pnp* type). The **pnp type** BJT is identical to the **npn type**, except all of the voltages and currents are in the reverse direction in the *pnp* type. The arrow on the transistor emitter indicates the transistor type by indicating the direction of the emitter current. In the *npn* type the arrow points away from the transistor, and in the *pnp* type the arrow points toward the transistor. In the next few experiments you will study the *npn* type BJT.

The bipolar junction transistor (BJT) **collector characteristic curves** can be plotted by recording the collector current (I_C) as a function of collector-emitter voltage (V_{CE}) for different base currents. The

BJT **base characteristic curve** can be plotted by recording the base-emitter voltage (V_{BE}) as a function of base current (I_B) with a constant collector-emitter voltage (V_{CE}). The BJT **emitter characteristic curve** can be plotted by recording the base-emitter voltage (V_{BE}) as a function of emitter current (I_E) with a constant collector-emitter voltage (V_{CE}). The circuit for plotting the BJT characteristic curves is shown in Figure 11-1.

The **saturation region** of the collector characteristic curves is that region where the collector-emitter voltage is below 0.7 V. In this region, the transistor is considered to be in the "on" state because it is close to being a short circuit. The base-emitter junction and the base-collector junction are both forward-biased when the transistor is in saturation. The **cutoff region** is the $I_B = 0$ line. At cutoff the transistor is considered to be in the "off" state because it is close to being an open circuit, with a very low collector current that is slightly above zero. This collector current is called **collector-emitter leakage current (I_{CEO})**. When the transistor is at cutoff, the base-emitter junction and the base-collector junction are both reverse-biased. The **active region** is between the cutoff region and the saturation region, where the collector-emitter voltage is above 0.7 V. In this region the base-emitter junction is forward-biased and the base-collector junction is reverse-biased. The collector current (I_C) is dependent on the value of the base current (I_B), but is not dependent on the value of the collector-emitter voltage (V_{CE}) in this region. Therefore, a BJT is a **current-controlled device**. In the active region, the collector current (I_C) can be represented as

$$I_C = \beta_{DC}I_B + I_{CEO}$$

and

$$\beta_{DC} = \frac{I_C - I_{CEO}}{I_B} \approx \frac{I_C}{I_B}$$

where β_{DC} is called the **dc current gain** of the transistor. The dc current gain (β_{DC}) is not constant, but varies slightly with changes in dc collector current.

In a bipolar junction transistor (BJT), the emitter current (I_E) is equal to the sum of the base current (I_B) and the collector current (I_C). Therefore,

$$I_E = I_B + I_C$$

The collector current is normally much larger than the base current. Therefore, the emitter current is approximately equal to the collector current in most cases.

The **ac current gain (β_{ac})** can be calculated from the collector characteristic curves at a particular value of V_{CE} by measuring the change in collector current (ΔI_C) due to a change in base current (ΔI_B), keeping the collector-emitter voltage (V_{CE}) constant. Therefore,

$$\beta_{ac} = \frac{\Delta I_C}{\Delta I_B} = \frac{I_{C2} - I_{C1}}{I_{B2} - I_{B1}}$$

The **ac input (base) resistance (r_{in})** can be calculated from the input (base) characteristic curve plotted at a particular collector-emitter voltage (V_{CE}) by measuring the change in base-emitter voltage (ΔV_{BE}) due to a change in base current (ΔI_B) at a particular point on the curve. Therefore,

$$r_{in} = \frac{\Delta V_{BE}}{\Delta I_B} = \frac{V_{BE2} - V_{BE1}}{I_{B2} - I_{B1}}$$

This makes r_{in} equal to the inverse of the slope of the tangent to the input (base) characteristic curve at that operating point.

The **ac emitter resistance (r_e)** can be calculated from the emitter characteristic curve plotted at a particular collector-emitter voltage (V_{CE}) by measuring the change in base-emitter voltage (ΔV_{BE}) due to a change in emitter current (ΔI_E) at a particular point on the curve. Therefore,

$$r_e = \frac{\Delta V_{BE}}{\Delta I_E} = \frac{V_{BE2} - V_{BE1}}{I_{E2} - I_{E1}}$$

This makes r_e equal to the inverse of the slope of the tangent to the emitter characteristic curve at that operating point. The ac emitter resistance can also be estimated from the formula

$$r_e \cong \frac{25mV}{I_E(mA)}$$

where I_E is the dc emitter current in mA.

Figure 11-1 Transistor Curve Plotter Circuit

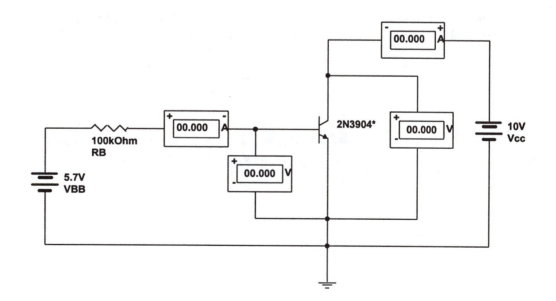

Procedure:

Step 1.   Open circuit file FIG11-1 and run the simulation. Record the readings for base current
              (I_B), collector current (I_C), and collector-emitter voltage (V_{CE}). Calculate the dc current
              gain (β_{DC}) based on the readings.

              $I_B =$ _____ $I_C =$ _____ $V_{CE} =$ _____

Step 2.   Change voltage source V_{BB} to 2.68 V, then run the simulation. Record the readings for
              base current (I_B), collector current (I_C), and collector-emitter voltage (V_{CE}). Calculate the
              dc current gain (β_{DC}) based on the readings.

              $I_B =$ _____ $I_C =$ _____ $V_{CE} =$ _____

Question: Based on the data collected in Steps 1 and 2, is the dc current gain (β_{DC}) constant or does it
vary when the collector current changes? Is the variation in β_{DC} large or small? **Explain.**

Step 3.   Change voltage source V_{CC} to 5 V, then run the simulation. Record the readings for base
              current (I_B), collector current (I_C), and collector-emitter voltage (V_{CE}).

              $I_B =$ _____ $I_C =$ _____ $V_{CE} =$ _____

Question: Based on the data collected in Steps 1-3, which is the collector current (I_C) most dependent
upon, the base current (I_B) or collector-emitter voltage (V_{CE})? **Explain.**

Step 4. Return voltage source V_{CC} to 10 V. Change voltage source V_{BB} to 0 V, then run the simulation. Record the readings for base current (I_B), collector current (I_C), and collector-emitter voltage (V_{CE}).

$I_B =$ _____ $I_C =$ _____ $V_{CE} =$ _____

Question: What is collector-emitter cutoff leakage current (I_{CEO})? What is the value of I_{CEO} for this 2N3904 transistor?

Step 5. Run the simulation for each value of $V_{BB}(I_B)$ and V_{CC} (V_{CE}) in Table 11-1. Record the values for collector current (I_C).

Table 11-1 Collector Current, I_C, in mA

$V_{BB}(V)$	$I_B(\mu A)$	V_{CE} in Volts					
		0.1	0.5	1	5	10	20
1.68	10						
2.68	20						
3.69	30						
4.70	40						
5.70	50						

Step 6. With V_{CC} (V_{CE}) set at 10 V, run the simulation for each value of V_{BB} (I_B) in Table 11-2. Record the values for the base-emitter voltage (V_{BE}) and the collector current (I_C). Note that the collector current (I_C) is approximately equal to the emitter current (I_E).

Table 11-2

$V_{BB}(V)$	$I_B(\mu A)$	$V_{BE}(V)$	$I_C{\approx}I_E(mA)$
0.70	1		
1.15	5		
1.68	10		
2.68	20		
3.69	30		
5.70	50		

Step 7.   From the values in Table 11-1 in Step 5, plot the I_C and V_{CE} data points on the graph for each value of I_B. Then draw the curve plot for each base current (I_B).

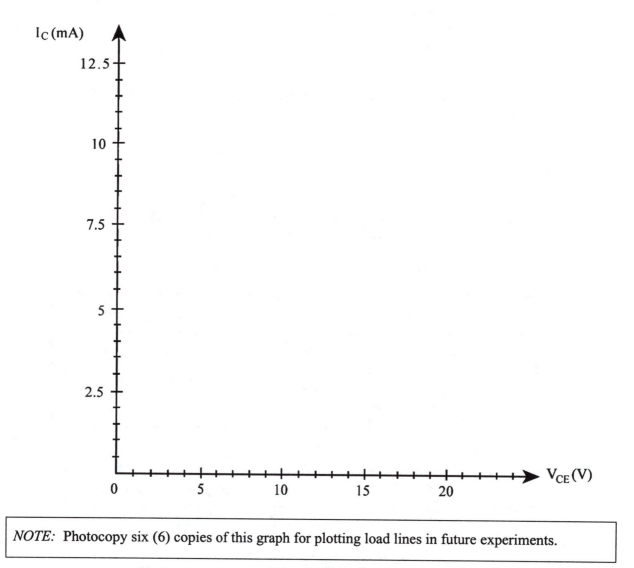

NOTE: Photocopy six (6) copies of this graph for plotting load lines in future experiments.

Question: What does the active region of the collector characteristics tell you about the dependency of collector current (I_C) on base current (I_B)? On collector-emitter voltage (V_{CE})?

Step 8.   From the values in Table 11-2 in Step 6, plot the I_B and V_{BE} data points on the graph. Then draw the curve plot.

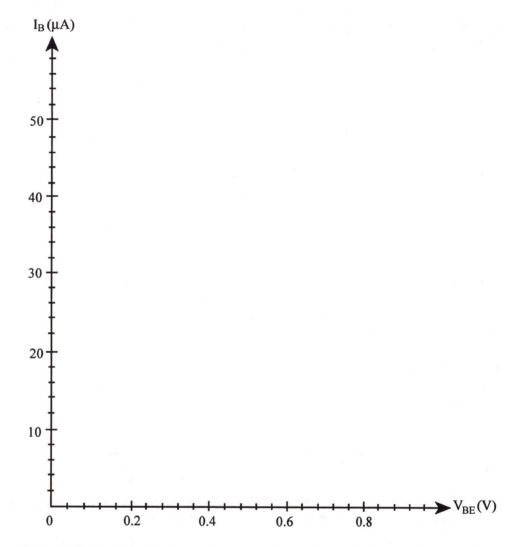

Question: What does the base characteristic curve tell you about the base-emitter junction compared to a forward-biased diode?

Step 9.   From the values in Table 11-2 in Step 6, plot the I_E and V_{BE} data points on the graph. Then draw the curve plot.

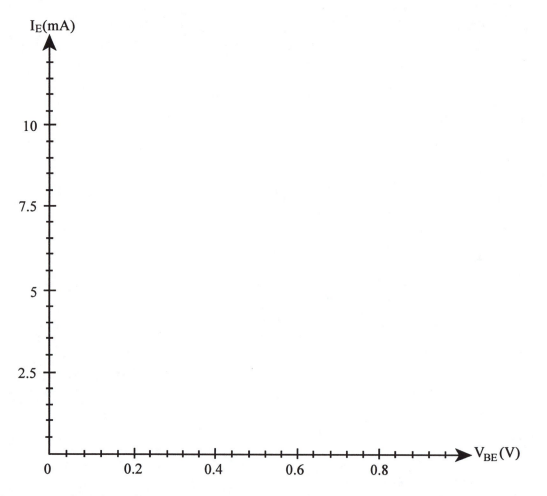

Step 10.   Based on the collector characteristics plotted in Step 7, calculate the ac current gain (β_{ac}) as the base current changes from 10 μA to 30 μA at $V_{CE} = 10$ V.

Step 11.   Based on the curve plot in Step 8, calculate the ac input resistance (r_{in}) from the change in base-emitter voltage as the base current changes from 10 μA to 30 μA.

Question: Is the ac input resistance the same at all points on the base characteristic curve? **Explain.**

Step 12.   Based on the curve plot in Step 9, calculate the ac emitter resistance (r_e) for a change in base-emitter voltage and emitter current as the base current changes from 10 μA to 30 μA.

Question: Is r_e the same for all values of emitter current? **Explain.**

Step 13.   Calculate the value of the ac emitter resistance (r_e) from the formula $r_e \cong 25mV/I_E(mA)$ using the value for I_E at $I_B = 20$ μA from Table 11-2.

Question: How does the value of measured emitter resistance (r_e) compare with the value calculated from the formula $r_e \cong 25mV/I_E(mA)$ using the value for I_E at $I_B = 20$ μA?

Name_____

Date_____

12

Base-Biased *NPN* Transistor

Objectives:

1. Draw the dc load line on the collector characteristics for a base-biased *npn* common emitter transistor configuration.
2. Locate the operating point (Q-point) on the dc load line.
3. Calculate the expected base current for a base-biased *npn* common emitter transistor configuration and compare the calculated value with the measured value.
4. Calculate the expected collector current for a base-biased *npn* common emitter transistor configuration and compare the calculated value with the measured value.
5. Calculate the expected collector-emitter voltage for a base-biased *npn* common emitter transistor configuration and compare the calculated value with the measured value.
6. Determine bias stability for a base-biased *npn* common emitter transistor configuration.
7. Determine the value of base resistance needed to drive the transistor into saturation.
8. Determine what needs to be done to the base circuit to drive the transistor into cutoff.

Materials:

One 2N3904 bipolar junction transistor
One 20 V dc power supply
Two 0–20 V dc voltmeters
One 0–10 mA dc milliammeter
One 0–100 μA dc microammeter
Resistors: one 2 kΩ, one 960 kΩ

Theory:

There are three ways to connect a bipolar junction transistor (BJT), **common emitter (CE)**, **common base (CB)**, and **common collector (CC)**. In the common emitter connection the emitter is at ground potential, in the common base connection the base is at ground potential, and in the common collector connection the collector is at ground potential. The common emitter connection is the most commonly used and will be used in this experiment. An example of a common emitter connection is shown in Figure 12-1.

A bipolar junction transistor (BJT) must be **dc biased** in the **active region** of the collector characteristics in order to operate as an amplifier. DC biasing establishes a steady level of base current (I_B), collector current (I_C), and collector-emitter voltage (V_{CE}), called the **dc operating point (Q-point)**. The dc operating point (Q-point) is selected so that the signal levels at the input are accurately reproduced at the output without distortion. When the dc operating point is in the active region of the collector characteristics, the transistor base-emitter junction is forward-biased and the base-collector junction is reverse-biased.

There are four methods of biasing a BJT common emitter configuration in the active region of the collector characteristics: base biased, voltage-divider biased, emitter biased, and collector-feedback biased. The circuit in Figure 12-1 is a **base-biased *npn* common emitter configuration**. The major disadvantage of a base-biased configuration is that the operating point (Q-point) is dependent on the value of the transistor dc current gain (β_{DC}), which is temperature dependent. Also, there is a large variation in dc current gain (β_{DC}) from one transistor to another of the same type. This means that the operating point (Q-point) will be unstable with base biasing.

The **dc load line** is a line on the collector characteristic curve plot that represents all of the possible dc operating points based on the circuit component values external to the transistor. The slope of the dc load line depends on the value of the external collector resistance. The dc load line can be located by determining the horizontal and vertical axis crossing points. The **horizontal axis crossing point** represents the value of the collector-emitter voltage (V_{CE}) when the collector current is zero ($I_C = 0$). For the base-biased *npn* common-emitter transistor circuit in Figure 12-1, the dc load line crosses the horizontal axis on the collector characteristic curve plot at $V_{CE} = V_{CC}$. The **vertical axis crossing point** represents the value of the collector current (I_C) when the transistor collector-emitter voltage is zero ($V_{CE} = 0$). For the circuit in Figure 12-1, the dc load line crosses the vertical axis on the collector characteristic curve plot at $I_C = V_{CC}/R_C$.

The operating point (Q-point) is located on the dc load line from the value of the collector current (I_C) or the value of the collector-emitter voltage (V_{CE}). Keeping the operating point (Q-point) as close as possible to the center of the load line allows for operating point drift without getting too close to cutoff or saturation of the transistor. This reduces the possibility of signal distortion.

The **base current (I_B)** for the base-biased transistor circuit in Figure 12-1 is calculated by finding the current in the base resistor, R_B. Therefore,

$$I_B = \frac{V_{CC} - V_{BE}}{R_B}$$

The **collector current (I_C)** is calculated by multiplying the base current by the dc current gain (β_{DC}), neglecting the collector leakage current (I_{CEO}). Therefore,

$$I_C = \beta_{DC}I_B + I_{CEO} \cong \beta_{DC}I_B$$

The **collector-emitter voltage (V_{CE})** is calculated from **Kirchhoff's voltage law**. Therefore,

$$V_{CC} = V_{CE} + I_C R_C$$
and
$$V_{CE} = V_{CC} - I_C R_C$$

In order to calculate the value of R_B needed to drive the transistor into saturation, you must first calculate the value of **collector current at saturation ($I_{C(sat)}$)**. The collector current at saturation is found from

$$I_{C(sat)} \cong \frac{V_{CC}}{R_C} \quad (V_{CE} \cong 0)$$

Next, you must calculate the **base current at saturation ($I_{B(sat)}$)**. The base current at saturation can be found from

$$I_{B(sat)} = \frac{I_{C(sat)}}{\beta_{DC}}$$

Based on the value of $I_{B(sat)}$, you can calculate the maximum value of the base resistance (R_B) from

$$R_B = \frac{V_{CC} - V_{BE}}{I_{B(sat)}} \cong \frac{V_{CC}}{I_{B(sat)}}$$

Figure 12-1 Base-Biased *NPN* Common Emitter Circuit

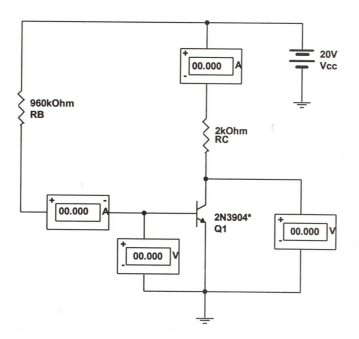

Procedure:

Step 1.   Open circuit file FIG12-1 and run the simulation. Record the collector current (I_C), base current (I_B), collector-emitter voltage (V_{CE}), and base-emitter voltage (V_{BE}).

$I_C =$ _____ $I_B =$ _____

$V_{CE} =$ _____ $V_{BE} =$ _____

Step 2.   For the circuit in Figure 12-1, draw the dc load line on one of the copies you made of the 2N3904 collector characteristics plotted in Step 7, Experiment 11. Based on the current and voltage readings in Step 1 above, locate the Q-point on the load line.

Question: Was the operating point (Q-point) close to the center of the load line? What advantage is this?

Step 3.   For the circuit in Figure 12-1, calculate the expected base current (I_B).

Question: How did the calculated base current (I_B) compare with the measured value?

Step 4.   For the circuit in Figure 12-1, calculate the expected collector current (I_C) using the value of I_B measured in Step 1 in this experiment and the value of β_{DC} measured in Step 2, Experiment 11.

Question: How did the calculated value for the collector current (I_C) compare with the measured value?

Step 5.   For the circuit in Figure 12-1, calculate the expected collector-emitter voltage (V_{CE}).

Question: How did the calculated value for the collector-emitter voltage (V_{CE}) compare with the measured value?

Step 6.   Double click transistor Q1, and then click *Edit Model*. Change the forward current gain coefficient (Bf) from 800 to 200, and then click *Change Part Model*. Click *OK* on the Models window to return to the circuit. You have changed the current gain (β) for the 2N3904 transistor. This will allow you to examine the effect that changing transistors would have on the collector current (Q-point) for a base-biased circuit. Run the simulation and record the collector current (I_C), base current (I_B), and collector-emitter voltage (V_{CE}).

$I_C =$ _____ $I_B =$ _____ $V_{CE} =$ _____

NOTE: If you are using this manual in a lab environment, you should change the 2N3904 transistor in place of changing the current gain (β).

Step 7.   Based on the new values for V_{CE} and I_C, locate the new Q-point on the load line drawn in Step 2. The characteristic curves no longer have the correct base currents because changing β moved the curves.

Question: Was there much change in the location of the operating point (Q-point) when the current gain (β) was changed? What disadvantage is this biasing method for a transistor amplifier?

Step 8.   Double click transistor Q1, click *Edit Model*, and return the value of forward current gain coefficient (Bf) back to 800 for the 2N3904 transistor. Don't forget to click *Change Part Model* and *OK*.

NOTE: If you are using this manual in a lab environment, you should change back to the original 2N3904 transistor.

Step 9.   Calculate the value of R_B needed to drive the transistor into saturation. Use the value of β_{DC} determined in Step 2, Experiment 11.

Step 10.   Change R_B to a value less than the value calculated in Step 9, then run the simulation. Record the value of the collector current (I_C), base current (I_B), and collector-emitter voltage (V_{CE}).

$I_C =$ _____ $I_B =$ _____ $V_{CE} =$ _____

Step 11.   Reduce the value of R_B significantly and run the simulation again. If the transistor is in saturation, there should be very little change in collector current (I_C) from the value in Step 10, even though there is a big change in the base current (I_B).

Question: Did you calculate the correct value for R_B to drive the transistor into saturation (little change in collector current when the base current changes)?

Step 12.   Make whatever change to the base circuit that is needed to drive the transistor into cutoff, then run the simulation. Record the value of the collector current (I_C) and collector-emitter voltage (V_{CE}).

$I_C =$ _____ $V_{CE} =$ _____

Questions: Is the transistor in cutoff?

What did you need to do to the base circuit to drive the transistor into cutoff?

Troubleshooting Problems

1.   Open circuit file FIG12-2 and run the simulation. Based on the measured voltages and currents, what is wrong with transistor Q1?

2.   Open circuit file FIG12-3 and run the simulation. Based on the measured voltages and currents, what is wrong with transistor Q1?

3.   Open circuit file FIG12-4 and run the simulation. Determine which circuit component is defective and state the defect (open or short). You can make any measurement desired.

Defective component: _____   Defect: _____

Name_____

Date_____

13

Voltage-Divider Biased *NPN* Transistor

Objectives:

1. Draw the dc load line on the transistor collector characteristics for a voltage-divider biased *npn* common emitter transistor configuration.
2. Locate the operating point (Q-point) on the dc load line.
3. Calculate the expected base voltage for a voltage-divider biased *npn* common emitter transistor configuration and compare the calculated value with the measured value.
4. Calculate the expected emitter current and collector current for a voltage-divider biased *npn* common emitter transistor configuration and compare the calculated values with the measured values.
5. Calculate the expected collector-emitter voltage for a voltage-divider biased *npn* common emitter transistor configuration and compare the calculated value with the measured value.
6. Calculate the dc current gain for the transistor based on the current readings.
7. Determine bias stability for a voltage-divider biased *npn* common emitter transistor configuration.

Materials:

One 2N3904 bipolar junction transistor
One 20 V dc power supply
Two 0–10 V dc voltmeters
Two 0–10 mA dc milliammeters
One 0–50 μA dc microammeter
Resistors: one 660 Ω, two 2 kΩ, one 10 kΩ

Theory:

Before attempting Experiment 13, you should complete Experiments 11 and 12.

Voltage-divider biasing, as shown in Figure 13-1, is the most widely used method of biasing because it produces a very **stable operating point (Q-point)**. With voltage-divider biasing, a dc voltage is applied to the transistor base by a voltage-divider network. If the transistor base current is much less than the voltage-divider network current, the base voltage will be very stable even with large variations in dc base current. This stable dc base voltage forces the dc emitter voltage (V_E) to be stable. Because the emitter current (I_E) and the collector current (I_C) depend on the emitter voltage, they will also be stable and have very little variation with variations in transistor current gain (β_{DC}). By reducing the variation in collector current, the operating point (Q-point) will be more stable than with other biasing methods.

For the **voltage-divider biased *npn* common-emitter circuit** shown in Figure 13-1, the dc load line crosses the horizontal axis on the collector characteristic curve plot at a value of collector-emitter voltage (V_{CE}) equal to V_{CC} ($I_C = 0$). The dc load line crosses the vertical axis at a value of collector current (I_C) equal to $V_{CC}/(R_C + R_E)$ ($V_{CE} = 0$).

The **operating point (Q-point)** is located on the dc load line. It is determined from the value of the collector current (I_C) or the value of the collector-emitter voltage (V_{CE}).

The **base voltage (V_B)** for the voltage-divider biased circuit in Figure 13-1 is calculated using the voltage divider rule. If $\beta(R_E) \gg R_2$, the transistor base current will be much less than the voltage-divider current and

$$V_B \cong \frac{V_{CC}R_2}{R_1 + R_2}$$

The **emitter current (I_E)** for the voltage-divider biased circuit in Figure 13-1 is calculated by first determining the emitter voltage (V_E). The emitter voltage is determined by subtracting the base-emitter voltage (V_{BE}) from the base voltage (V_B). Then the emitter current is calculated by dividing the emitter voltage by the emitter resistor value (R_E). Therefore,

$$I_E = \frac{V_E}{R_E} = \frac{V_B - V_{BE}}{R_E}$$

The **collector current (I_C)** can be estimated from the emitter current as follows:

$$I_C = I_E - I_B \cong I_E$$

The **collector-emitter voltage (V_{CE})** for the voltage-divider biased circuit in Figure 13-1 is calculated from **Kirchhoff's voltage law**. Therefore,

$$V_{CC} = I_C R_C + V_{CE} + I_E R_E$$

and

$$V_{CE} = V_{CC} - I_C R_C - I_E R_E$$

The **dc current gain (β_{DC})** is calculated by dividing the dc collector current (I_C) by the dc base current (I_B). Therefore, neglecting leakage current,

$$\beta_{DC} \cong \frac{I_C}{I_B}$$

Figure 13-1 Voltage-Divider Biased *NPN* Common Emitter Circuit

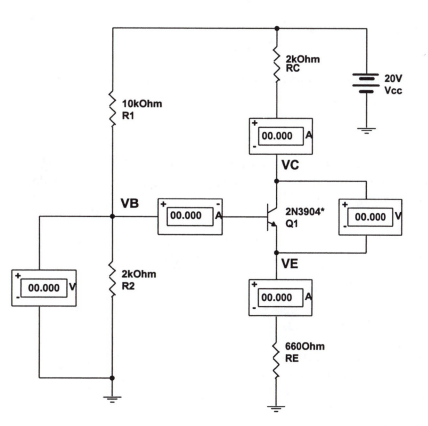

Procedure:

Step 1.   Open circuit file FIG13-1 and run the simulation. Record the collector current (I_C), emitter current (I_E), base current (I_B), collector-emitter voltage (V_{CE}), and base voltage (V_B).

$I_C =$ _____ $I_E =$ _____ $I_B =$ _____

$V_{CE} =$ _____ $V_B =$ _____

Step 2.   For the circuit in Figure 13-1, draw the dc load line on a copy you made of the 2N3904 collector characteristics plotted in Step 7, Experiment 11. Based on the current and voltage readings in Step 1 above, locate the Q-point on the load line.

Question: Was the operating point (Q-point) close to the center of the load line? What advantage is this?

Step 3.   For the circuit in Figure 13-1, calculate the expected dc base voltage (V_B).

Question: How did the calculated value for the base voltage (V_B) compare with the measured value?

Step 4.   Based on the base voltage calculated in Step 3, calculate the expected dc emitter current (I_E) and estimate the collector current (I_C) from the emitter current value. (Estimate V_{BE} to be 0.7 V).

Question: How did the calculated value for the emitter current (I_E) compare with the measured value? What is the relationship between emitter current (I_E) and collector current (I_C)?

Step 5.   For the circuit in Figure 13-1, calculate the expected collector-emitter voltage (V_{CE}) based on the value of I_C and I_E calculated in Step 4.

Question: How did the calculated value for the collector-emitter voltage (V_{CE}) compare with the measured value?

Step 6.   Calculate the dc current gain (β_{DC}) for the transistor, based on the current readings in Step 1.

Question: How did the calculated value for the dc current gain (β_{DC}) compare with the value measured in Experiment 11, Step 2?

Step 7.   Double click transistor Q1, and then click *Edit Model*. Change the forward current gain coefficient (Bf) from 800 to 200, and then click *Change Part Model*. Click *OK* on the Models window to return to the circuit. You have changed the current gain (β) for the 2N3904 transistor. This will allow you to examine the effect that changing transistors would have on the collector current (Q-point) for a voltage-divider biased circuit. Run the simulation and record the collector current (I_C), base current (I_B), and collector-emitter voltage (V_{CE}).

$I_C =$ _____ $I_B =$ _____ $V_{CE} =$ _____

NOTE: If you are using this manual in a lab environment, you should change the original 2N3904 transistor in place of changing the current gain (β).

Step 8.   Based on the new values for V_{CE} and I_C, locate the new Q-point on the load line drawn in Step 2. (The characteristic curves no longer have the correct base currents because changing β moved the curves).

Questions: Was there much change in the location of the operating point (Q-point) when the current gain (β) was changed? How did voltage-divider biasing compare with base biasing? What advantage is this biasing method over base bias?

Step 9.   Double-click transistor Q1 and return the value of forward current gain coefficient (Bf) back to 800 for the 2N3904 transistor. Don't forget to click *Edit Model, Change Part Model* and *OK*.

NOTE: If you are using this manual in a lab environment, you should change back to the original 2N3904 transistor.

Troubleshooting Problems

1.   Open circuit file FIG13-2 and run the simulation. Based on the measured voltages and currents, what is wrong with transistor Q1?

2.   Open circuit file FIG13-3 and run the simulation. Based on the measured voltages and currents, what is wrong with transistor Q1?

3.   Open circuit file FIG13-4 and run the simulation. Determine which circuit component is defective and state the defect (open or short). You can make any measurement desired.

  Defective component: _____   Defect: _____

4.   Open circuit file FIG13-5 and run the simulation. Determine which circuit component is defective and state the defect (open or short). You can make any measurement desired.

  Defective component: _____   Defect: _____

5.   Open circuit file FIG13-6 and run the simulation. Determine which circuit component is defective and state the defect (open or short). You can make any measurement desired.

  Defective component: _____   Defect: _____

Name_____

Date_____

14

Voltage-Divider Biased *PNP* Transistor

Objectives:

1. Calculate the expected base voltage for a voltage-divider biased *pnp* transistor and compare the calculated value with the measured value.
2. Calculate the expected emitter current and collector current for a voltage-divider biased *pnp* transistor and compare the calculated values with the measured values.
3. Calculate the expected emitter-collector voltage for a voltage-divider biased *pnp* transistor and compare the calculated value with the measured value.
4. Calculate the dc current gain (β_{DC}) for a voltage-divider biased *pnp* transistor based on the current readings.
5. Determine whether the operating point (Q-point) is close to the middle of the load line based on the current and voltage readings.
6. Determine bias stability for a voltage-divider biased *pnp* transistor.

Materials:

One 2N3906 bipolar junction *pnp* transistor
One dc power supply
Two 0–10 V dc voltmeters
Two 0–10 mA dc milliammeters
One 0–50 μA dc microammeter
Resistors (one each): 680 Ω, 2 kΩ, 5 kΩ, 20 kΩ

Theory:

All of the voltages and currents are in the opposite direction in a ***pnp* bipolar transistor** compared to an *npn* bipolar transistor. Therefore, the *pnp* transistor requires bias voltage polarities opposite to the polarities for an *npn* transistor. This is usually accomplished by turning the transistor upside down, as shown in Figure 14-1. This allows the positive supply voltage (V_{EE}) to remain on top of the circuit diagram and ground to remain on the bottom. The analysis procedure is basically the same as the *npn* transistor in Experiment 13.

The **base voltage (V_B)** for the voltage-divider biased *pnp* transistor circuit shown in Figure 14-1 is calculated using the voltage divider rule. If $\beta(R_E) \rangle\rangle R_2$, the transistor base current will be much less than the voltage-divider current and

$$V_B \cong \frac{V_{EE}R_1}{R_1 + R_2}$$

The **emitter current (I_E)** is calculated by first determining the emitter voltage (V_E). The emitter voltage is determined by adding the emitter-base voltage (V_{EB}) to the base voltage (V_B). The emitter current is calculated by subtracting the emitter voltage from the supply voltage (V_{EE}) and then dividing by the emitter resistor value (R_E). Therefore,

$$I_E = \frac{V_{EE} - V_E}{R_E} = \frac{V_{EE} - (V_B + V_{EB})}{R_E}$$

The **collector current (I_C)** can be estimated from the emitter current as follows:

$$I_C = I_E - I_B \cong I_E$$

The **emitter-collector voltage (V_{EC})** for the voltage-divider biased *pnp* transistor circuit in Figure 14-1 is calculated from **Kirchhoff's voltage law**. Therefore,

$$V_{EE} = I_C R_C + V_{EC} + I_E R_E$$

and

$$V_{EC} = V_{EE} - I_C R_C - I_E R_E$$

The **dc current gain (β_{DC})** is calculated by dividing the dc collector current (I_C) by the dc base current (I_B). Therefore, neglecting leakage current,

$$\beta_{DC} \cong \frac{I_C}{I_B}$$

Figure 14-1 Voltage-Divider Biased *PNP* Transistor Circuit

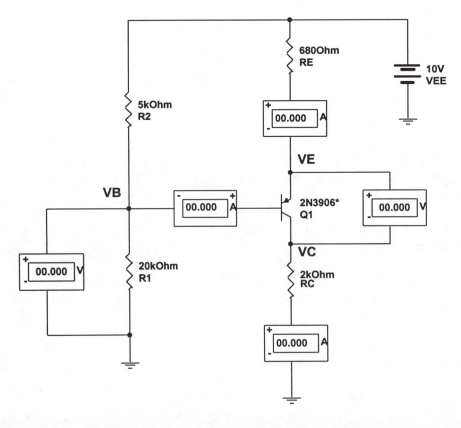

Procedure:

Step 1.   Open circuit file FIG14-1 and run the simulation. Record the collector current (I_C), emitter current (I_E), base current (I_B), emitter-collector voltage (V_{EC}), and base voltage (V_B).

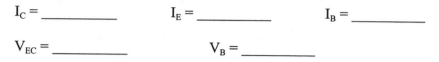

$I_C =$ _____ $I_E =$ _____ $I_B =$ _____

$V_{EC} =$ _____ $V_B =$ _____

Question: Based on the current and voltage readings in Step 1, was the operating point (Q-point) close to the center of the load line? How did you determine this?

Step 2.   For the circuit in Figure 14-1, calculate the expected base voltage (V_B).

Question: How did the calculated value for the base voltage (V_B) compare with the measured value?

Step 3.   Based on the base voltage determined in Step 2, calculate the expected emitter current (I_E) and estimate the collector current (I_C) from the emitter current value. (Assume V_{EB} to be 0.7 V).

Question: How did the calculated value for the emitter current (I_E) compare with the measured value? What is the relationship between emitter current (I_E) and collector current (I_C)?

Step 4.   For the circuit in Figure 14-1, calculate the expected emitter-collector voltage (V_{EC}) based on the values for I_C and I_E calculated in Step 3.

Question: How did the calculated value for the emitter-collector voltage (V_{EC}) compare with the measured value?

Step 5.   For the circuit in Figure 14-1, calculate the current gain (β) for the 2N3906 transistor, based on the current readings in Step 1.

Question: How did this value compare with the value for the forward current gain (β) for the 2N3905 measured in Experiment 11?

Step 6.   Double click transistor Q1, and then click *Edit Model*. Change the forward current gain coefficient (Bf) from 200 to 400, and then click *Change Part Model*. Click *OK* on the Models window to return to the circuit. You have changed the current gain (β) for the 2N3906 transistor. This will allow you to examine the effect that changing transistors would have on the collector current (Q-point) for a voltage-divider biased circuit. Run the simulation and record the collector current (I_C), base current (I_B), and emitter-collector voltage (V_{EC}).

$I_C =$ _____ $I_B =$ _____ $V_{EC} =$ _____

NOTE: If you are using this manual in a lab environment, you should change the 2N3906 transistor in place of changing the current gain (β).

Question: Based on the current and voltage readings in Steps 1 and 6, was there much change in the collector current (I_C) and the emitter-collector voltage (V_{EC})(Q-point) when the current gain (β) was changed? Were these results the same as the results for the *npn* transistor? **Explain.**

Step 7.   Double click transistor Q1 and return the value of forward current gain (Bf) back to 200 for the 2N3906 transistor. Don't forget to click *Edit Model*, *Change Part Model*, and *OK*.

NOTE: If you are using this manual in a lab environment, you should change back to the original 2N3906 transistor.

Troubleshooting Problems

1.   Open circuit file FIG14-2 and run the simulation. Based on the measured voltages and currents, what is wrong with transistor Q1?

2.   Open circuit file FIG14-3 and run the simulation. Based on the measured voltages and currents, what is wrong with transistor Q1?

3.   Open circuit file FIG14-4 and run the simulation. Determine which component is defective and state the defect (open or short).

        Defective component: _____   Defect: _____

15

Emitter Bias

Objectives:

1. Draw the dc load line on the collector characteristics for an emitter-biased *npn* transistor configuration.
2. Locate the operating point (Q-point) on the dc load line.
3. Calculate the base voltage for an emitter-biased *npn* transistor configuration and compare the calculated value with the measured value.
4. Calculate the expected emitter current and collector current for an emitter-biased *npn* transistor configuration and compare the calculated values with the measured values.
5. Calculate the expected collector-emitter voltage for an emitter-biased *npn* transistor configuration and compare the calculated value with the measured value.
6. Calculate the dc current gain (β_{DC}) for the transistor based on the current readings.
7. Determine bias stability for an emitter-biased *npn* transistor configuration.

Materials:

One 2N3904 bipolar junction transistor
Two 0–20 V dc power supplies
Two 0–20 V dc voltmeters
Two 0–10 mA dc milliammeters
One 0–50 µA dc microammeter
Resistors (one each): 500Ω , 1.5 kΩ, 100 kΩ

Theory:

Emitter biasing, as shown in Figure 15-1, requires **two separate dc voltage sources**, one positive and one negative. Voltage source V_{CC} is the positive dc source and voltage source V_{EE} is the negative dc source. Kirchhoff's voltage law applied to the base-emitter circuit gives the equation

$$V_{EE} = I_B R_B + V_{BE} + I_E R_E$$

The **base voltage (V_B)** for the emitter-biased transistor circuit is negative and is equal to the voltage across the base resistor (R_B) due to the base current (I_B). Therefore,

$$V_B = -I_B R_B$$

and

$$V_{EE} = -V_B + V_{BE} + I_E R_E$$

Solving for the **emitter current (I_E)** gives the equation

$$I_E = \frac{V_{EE} + V_B - V_{BE}}{R_E}$$

Note: V_B is negative

Notice that if the base-emitter voltage (V_{BE}) and the base voltage (V_B) are much less than the voltage of voltage source V_{EE}, then the emitter current is practically dependent upon V_{EE} and R_E. Therefore,

$$I_E \cong \frac{V_{EE}}{R_E}$$

and the transistor parameters have very little effect on the emitter current. The **collector current (I_C)** can be estimated from the emitter current (I_E) as follows:

$$I_C = I_E - I_B \cong I_E$$

This means that the transistor parameters have very little effect on collector current also. This makes the operating point (Q-point) very stable with emitter biasing.

The **collector-emitter voltage (V_{CE})** for the emitter-biased *npn* transistor circuit in Figure 15-1 is calculated from **Kirchhoff's voltage law**. Therefore,

$$V_{CC} + V_{EE} = I_C R_C + V_{CE} + I_E R_E$$

and

$$V_{CE} = V_{CC} + V_{EE} - I_C R_C - I_E R_E$$

For the emitter-biased circuit in Figure 15-1, the **dc load line** crosses the horizontal axis on the collector characteristic curve plot at a value of collector-emitter voltage (V_{CE}) equal to $V_{CC} + V_{EE}$ (I_C and I_E equal to zero). It crosses the vertical axis at a value of collector current (I_C) equal to $(V_{CC} + V_{EE})/(R_C + R_E)$ (V_{CE} equal to zero).

The **operating point (Q-point)** is located on the dc load line based on the value of the collector current (I_C) or the collector-emitter voltage (V_{CE}).

The **dc current gain (β_{DC})** is calculated by dividing the dc collector current (I_C) by the dc base current (I_B). Therefore, neglecting leakage current,

$$\beta_{DC} \cong \frac{I_C}{I_B}$$

Figure 15-1 Emitter-Biased *NPN* Transistor Circuit

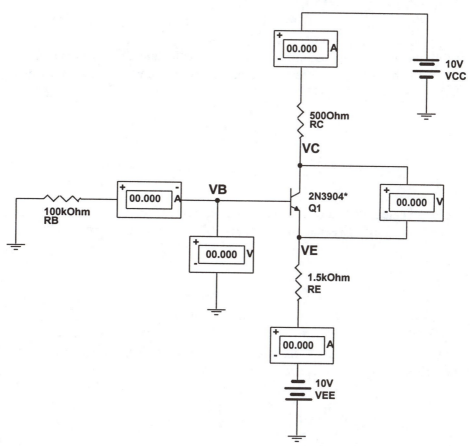

Procedure:

Step 1.   Open circuit file FIG15-1 and run the simulation. Record the collector current (I_C), emitter current (I_E) base current (I_B), collector-emitter voltage (V_{CE}), and base voltage (V_B).

$I_C =$ _____ $I_E =$ _____ $I_B =$ _____

$V_{CE} =$ _____ $V_B =$ _____

Step 2.   For the circuit in Figure 15-1, draw the dc load line on a clean copy of the 2N3904 collector characteristics plotted in Step 7, Experiment 11. Based on the current and voltage readings in Step 1 above, locate the Q-point on the load line.

Question: Was the operating point (Q-point) close to the center of the load line? What advantage is this?

Step 3.   Based on the base current (I_B) in the circuit in Figure 15-1, calculate the base voltage (V_B).

Question: How did the calculated value for the base voltage (V_B) compare with the measured value?

Step 4.   Based on the base voltage determined in Step 3, calculate the expected emitter current (I_E) and estimate the collector current (I_C) from the emitter current value. (Estimate V_{BE} to be 0.7 V.)

Questions: How did the calculated value for the emitter current (I_E) compare with the measured value? What is the relationship between emitter current (I_E) and collector current (I_C)?

Step 5.   For the circuit in Figure 15-1, calculate the expected collector-emitter voltage (V_{CE}) based on the values of I_C and I_E determined in Step 4.

Question: How did the calculated value for the collector-emitter voltage (V_{CE}) compare with the measured value?

Step 6.   Based on the current readings in Step 1, calculate the dc current gain (β_{DC}) for the transistor.

Question: How did the calculated value for the dc current gain (β_{DC}) for the 2N3904 compare with the value of the dc current gain (β_{DC}) determined in Experiment 11?

Step 7.   Double click transistor Q1, and then click *Edit Model*. Change the forward current gain coefficient (Bf) from 800 to 200, and then click *Change Part Model*. Click *OK* on the Models window to return to the circuit. You have changed the current gain (β) for the 2N3904 transistor. This will allow you to examine the effect that changing transistors would have on the collector current (Q-point) for an emitter biased circuit. Run the simulation and record the collector current (I_C), base current (I_B), and collector-emitter voltage (V_{CE}).

$I_C =$ _____ $I_B =$ _____ $V_{CE} =$ _____

NOTE: If you are using this manual in a lab environment, you should change the 2N3904 transistor in place of changing the current gain (β).

Step 8.   Based on the new values for V_{CE} and I_C, locate the new Q-point on the load line drawn in Step 2. (The characteristic curves no longer have the correct base currents because changing β moved the curves).

Questions: Was there much change in the location of the operating point (Q-point) when the current gain (β) was changed? How did emitter bias stability compare with base bias stability?

What is the advantage of emitter bias over base bias?

What is the advantage of voltage-divider bias over emitter bias?

Step 9.   Double click transistor Q1 and return the value of the forward current gain coefficient (Bf) back to 800 for the 2N3904 transistor. Don't forget to click *Edit Model, Change Part Model*, and *OK*.

NOTE: If you are using this manual in a lab environment, you should change back to the original 2N3904 transistor.

Troubleshooting Problems

1.   Open circuit file FIG15-2 and run the simulation. Based on the measured voltages and currents, what is wrong with transistor Q1?

2.   Open circuit file FIG15-3 and run the simulation. Based on the measured voltages and currents, what is wrong with transistor Q1?

3.   Open circuit file FIG15-4 and run the simulation. Determine which circuit component is defective and state the defect (short or open). You can use any instrument available to make any measurement desired.

         Defective component: _____   Defect: _____

4.   Open circuit file FIG15-5 and run the simulation. Determine which circuit component is defective and state the defect (short or open). You can use any instrument available to make any measurement desired.

         Defective component: _____   Defect: _____

16

Name_____

Date_____

Collector-Feedback Bias

Objectives:

1. Draw the dc load line on the transistor collector characteristics for a collector-feedback biased *npn* transistor configuration.
2. Locate the operating point (Q-point) on the dc load line.
3. Calculate the dc current gain (β_{DC}) for the transistor based on the collector current and base current values.
4. Calculate the expected collector current for a collector-feedback biased *npn* transistor configuration and compare the calculated value with the measured value.
5. Calculate the expected collector-emitter voltage for a collector-feedback biased *npn* transistor and compare the calculated value with the measured value.
6. Determine bias stability for a collector-feedback biased *npn* transistor configuration.

Materials:

One 2N3904 bipolar junction transistor
One 0–20 V dc power supply
One 0–20 V dc voltmeter
One 0–10 mA dc milliammeter
One 0–50 μA dc microammeter
Resistors: one 2 kΩ, one 400 kΩ

Theory:

Collector-feedback bias, as shown in Figure 16-1, uses **negative feedback** from the collector to the base to reduce the effect of transistor parameter variations in order to produce a more stable operating point (Q-point). The negative feedback is accomplished by connecting the base resistor (R_B) to the transistor collector rather than to V_{CC}, as in base bias. This causes the collector voltage (V_{CE}) to provide the dc bias for establishing the base current (I_B). Variations in the transistor parameters (β and V_{BE}) cause the collector current (I_C) to change. When the collector current changes, the voltage drop across the collector resistor (R_C) changes. This causes the collector voltage (V_{CE}) to change and readjust the base current, returning the collector current to approximately its original value. This produces a more stable operating point (Q-point).

For a collector-feedback biased *npn* transistor circuit, the **collector current (I_C)** is calculated from

$$I_C = \frac{V_{CC} - V_{BE}}{R_C + \dfrac{R_B}{\beta_{DC}}}$$

For the collector-feedback biased *npn* transistor circuit in Figure 16-1, the **dc load line** crosses the horizontal axis on the collector characteristic curve plot at a value of collector-emitter voltage (V_{CE}) equal to V_{CC} ($I_C = 0$). It crosses the vertical axis at a value of collector current (I_C) equal to V_{CC}/R_C ($V_{CE} = 0$).

The **operating point (Q-point)** is located on the dc load line. It is determined from the value of the collector current (I_C). For **maximum operating point stability** with collector-feedback bias, the value of R_B/β_{DC} must be **much less than** the value of R_C. In order for the operating point to be close to the **middle of the load line** with collector-feedback bias, the value of R_C must be approximately **equal to** R_B/β_{DC}. Therefore, keeping the operating point close to the center of the load line requires some sacrifice in operating point stability.

The **dc current gain (β_{DC})** is calculated by dividing the dc collector current (I_C) by the dc base current (I_B). Therefore, neglecting leakage current,

$$\beta_{DC} \cong \frac{I_C}{I_B}$$

The **collector-emitter voltage (V_{CE})** for the collector-feedback biased *npn* transistor circuit in Figure 16-1 is calculated from **Kirchhoff's voltage law**. Therefore,

$$V_{CC} = (I_C + I_B)R_C + V_{CE} \cong I_C R_C + V_{CE}$$

and

$$V_{CE} \cong V_{CC} - I_C R_C$$

Collector-feedback bias is one of the simplest biasing networks, but the operating point is not as stable as with some of the other biasing configurations if it is desired to keep the operating point in the center of the load line.

Figure 16-1 Collector-Feedback Biased *NPN* Transistor Circuit

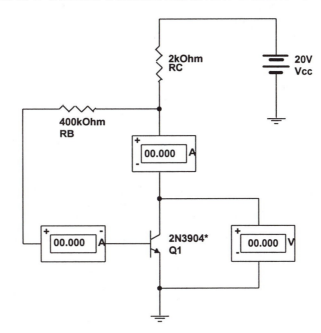

Procedure:

Step 1.   Open circuit file FIG6-1 and run the simulation. Record the collector current (I_C), base current (I_B), and collector-emitter voltage (V_{CE}).

$I_C = $ _____ $I_B = $ _____ $V_{CE} = $ _____

Step 2.   For the circuit in Figure 16-1, draw the dc load line on a copy of the 2N3904 collector characteristics plotted in Step 7, Experiment 11. Based on the current and voltage readings in Step 1 above, locate the Q-point on the load line.

Question: Was the operating point (Q-point) close to the center of the load line? What advantage is this?

Step 3.   From the values measured in Step 1, calculate the current gain (β_{DC}) for the 2N3904 transistor.

Question: How did the calculated value for the dc current gain (β_{DC}) compare with the dc current gain (β_{DC}) determined for the 2N3904 in Experiment 11?

Step 4.   For the circuit in Figure 16-1, calculate the expected collector current (I_C) using the value
                for β_{DC} calculated in Step 3. (Assume V_{BE} to be 0.7 V.)

Question: How did the calculated value for the collector current (I_C) compare with the measured
value?

Step 5.   For the circuit in Figure 16-1, calculate the expected collector-emitter voltage (V_{CE}) based
                on the value for I_C calculated in Step 4.

Question: How did the calculated value for the collector-emitter voltage (V_{CE}) compare with the
measured value?

Step 6.   Double click transistor Q1, and then click *Edit Model*. Change the forward current gain
                coefficient (Bf) from 800 to 200, and then click *Change Part Model*. Click *OK* on the
                Models window to return to the circuit. You have changed the current gain (β) for the
                2N3904 transistor. This will allow you to examine the effect that changing transistors
                would have on the collector current (Q-point) for a collector-feedback biased circuit. Run
                the simulation and record the collector current (I_C), base current (I_B), and collector-emitter
                voltage (V_{CE}).

                $I_C = $ _____ $I_B = $ _____ $V_{CE} = $ _____

NOTE: If you are using this manual in a lab environment, you should change the 2N3904 transistor
in place of changing the current gain (β).

Step 7.   Based on the new values for V_{CE} and I_C, locate the new Q-point on the load line drawn in
                Step 2. (Notice that the characteristic curves no longer have the correct base currents
                because changing β moved the curves).

Questions: Was there much change in the location of the operating point (Q-point) when the current
gain (β) was changed?

How did collector-feedback bias stability compare with base bias stability? With voltage-divider bias stability? With emitter bias stability?

Step 8.   Double click transistor Q1 and return the value of forward current gain coefficient (Bf) back to 800 for the 2N3904 transistor. Don't forget to click *Edit Model*, *Change Part Model*, and *OK*.

NOTE: If you are using this manual in a lab environment, you should change back to the original 2N3904 transistor.

Troubleshooting Problems

1.   Open circuit file FIG16-2 and run the simulation. Based on the measured voltages and currents, what is wrong with transistor Q1?

2.   Open circuit file FIG16-3 and run the simulation. Based on the measured voltages and currents, what is wrong with transistor Q1?

17

Small-Signal Common-Emitter Amplifier

Objectives:

1. Measure the voltage gain of a small-signal common-emitter (CE) amplifier and compare the measured value with the calculated value.
2. Determine the phase relationship between the input and output for a one-stage CE amplifier.
3. Measure the input resistance of a CE amplifier and compare the measured value with the calculated value.
4. Determine the effect of input source resistance on the overall voltage gain.
5. Measure the output resistance of a CE amplifier.
6. Determine the effect of load resistance on the overall voltage gain.
7. Observe the effect of coupling capacitance on the dc offset of the ac signal.
8. Determine the effect of unbypassed emitter resistance on the voltage gain for a CE amplifier.
9. Determine the effect of unbypassed emitter resistance on gain stability for a CE amplifier.

Materials:

One 2N3904 bipolar junction transistor
Capacitors: one 1.0 μF, one 10 μF, one 470 μF
One 20 V dc power supply
One dual-trace oscilloscope
One function generator
Resistors: one 660 Ω, one 1 kΩ, three 2 kΩ, one 10 kΩ, one 200 kΩ

Theory:

Experiments 11 and 13 should be completed before attempting this experiment.

The purpose of **biasing** a transistor amplifier circuit is to establish an **operating point (Q-point)** in the **active region** of the characteristic curves to cause linear amplifier response to ac input voltage variations. In applications where the ac input signal voltage variations are low, the collector current and collector-emitter voltage variations are relatively small. Therefore, the location of the operating point (Q-point) is not as critical as it would be for large input signal voltage variations, where it must be near the center of the load line. Amplifiers designed to handle low input voltage variations are called **small-signal amplifiers**.

The amplifier in Figure 17-1 is a **common-emitter small-signal amplifier** with voltage divider biasing. (See Experiment 13.) Capacitor C_1 is an **input coupling capacitor** that will prevent the dc base bias voltage from being affected by the input signal source. Capacitor C_2 is an **output coupling capacitor** that will prevent the dc collector bias voltage from being affected by the load resistance (R_L). These capacitor values need to be large enough to offer a very low reactance (X_C) relative to other circuit resistances at the lowest ac signal frequency expected. Capacitor C_3 is a **bypass capacitor** that should have a much lower reactance (X_C) than the transistor ac emitter resistance (r_e) at the lowest ac signal frequency. This will cause the emitter to be close to ground potential for the ac signal, but still maintain the dc emitter bias voltage. Without the emitter bypass capacitor (C_3), the total ac emitter resistance would be higher, causing a lower amplifier voltage gain (A_V).

In the amplifier in Figure 17-1, the ac signal source causes the dc base voltage to vary above and below the dc bias level, resulting in a base current variation. The base current variation will cause a much larger collector current variation due to the current gain (β) of the transistor. This variation in collector current will cause the operating point (Q-point) to move up and down the ac load line and cause a large variation in collector voltage. When the input voltage and current increase, the collector (output) voltage decreases, and when the input voltage and current decrease, the collector (output) voltage increases. This causes the **output voltage to be 180 degrees out-of-phase with the input voltage** for a **one-stage amplifier**. The ratio between the large collector (output) voltage variation and the small input voltage variation is the **amplifier voltage gain (A_V)**. Based on the ac peak-to-peak output and input voltages, the voltage gain of an amplifier is determined by dividing the ac peak-to-peak output voltage (V_o) by the ac peak-to-peak input voltage (V_{in}). Therefore,

$$A_V = \frac{V_o}{V_{in}}$$

The **expected voltage gain (A_V)** of the one-stage common-emitter amplifier shown in Figure 17-1 is calculated by dividing the ac collector resistance (R_c) by the transistor ac emitter resistance (r_e), determined in Step 12, Experiment 11. The ac collector resistance is equal to the parallel equivalent of the collector resistor (R_C) and the load resistor (R_L) . Therefore,

$$R_c = \frac{R_C R_L}{R_C + R_L}$$

and

$$A_V = \frac{R_c}{r_e}$$

For an amplifier with **unbypassed emitter resistance**, the voltage gain is

$$A_V = \frac{R_c}{r_e + R_E}$$

where R_E is the unbypassed emitter resistance.

The two voltage gain equations for bypassed and unbypassed emitter resistance demonstrate that bypassing R_E with a bypass capacitor (C_3) produces a higher gain, but there is more **gain stability** when R_E is unbypassed because the voltage gain is less dependent on the transistor emitter resistance (r_e). The larger the value of unbypassed emitter resistance (R_E) relative to the transistor emitter resistance (r_e), the more stable the voltage gain will be when the transistor parameters change, but the gain will be lower.

The **amplifier ac input resistance (R_{in})** can be determined by placing a 1 kΩ resistor between V_{in} and capacitor C_1 in Figure 17-1 and dividing the ac peak-to-peak input voltage on the base (V_b) by the ac peak-to-peak input current (I_{in}). Therefore,

$$R_{in} = \frac{V_b}{I_{in}}$$

where $I_{in} = (V_{in} - V_b)/1 \text{ k}\Omega$

The **expected amplifier ac input resistance (R_{in})** can be calculated from the circuit component values by first finding the transistor ac input resistance (r_i), and combining this resistance with the parallel equivalent resistance of the biasing network. The **transistor ac input resistance** is calculated from

$$r_i = (\beta_{ac})(r_e)$$

where r_e is the transistor ac emitter resistance determined in Step 12, Experiment 11 and β_{ac} is the ac current gain determined in Step 10, Experiment 11.

For unbypassed emitter resistance, the transistor ac input resistance becomes

$$r_i = (\beta_{ac})(r_e + R_E)$$

where R_E is the unbypassed emitter resistance.

The amplifier ac input resistance (R_{in}) is then found from the equation

$$\frac{1}{R_{in}} = \frac{1}{R_1} + \frac{1}{R_2} + \frac{1}{r_i}$$

by solving for R_{in}, where R_1 and R_2 are the biasing network resistors.

The **amplifier ac output resistance (R_o)** can be calculated from the ac peak-to-peak open circuit output voltage (V_{oc}) and the ac peak-to-peak output voltage (V_o) for a given load resistance (R_L) using Thevinin's theorem and solving the following equation for R_o:

$$\frac{V_o}{V_{oc}} = \frac{R_L}{R_o + R_L}$$

Note: Assume a load resistance of 200 kΩ or greater to be an open circuit.

Figure 17-1 Common-Emitter Amplifier

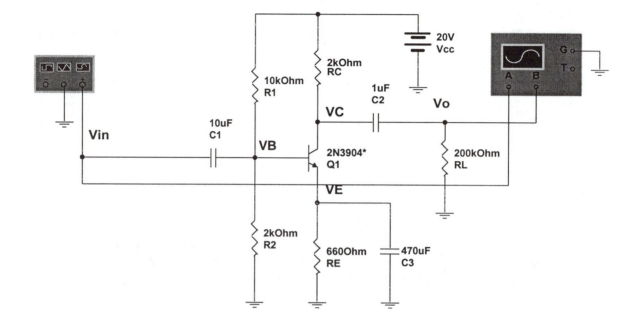

Procedure:

Step 1.   Open circuit file FIG17-1. Bring down the oscilloscope enlargement and make sure that the following settings are selected: Time base (Scale = 100 μs/Div, Xpos = 0, Y/T), Ch A (Scale = 1 mV/Div, Ypos = 0, AC), Ch B (Scale = 200 mV/Div, Ypos = 0, AC), Trigger (Pos edge, Level = 0 V, Sing, B). Bring down the function generator enlargement and make sure that the following settings are selected: *Sine Wave*, Freq = 2 kHz, Ampl = 2 mV, Offset = 0. Refer to the data collected in Experiment 13 for dc biasing currents and voltages. (If Experiment 13 has not been completed, do so before continuing).

Step 2.   Run the simulation. After one screen display, pause the simulation and record the ac peak-to-peak input voltage (V_{in}) and the ac peak-to-peak output voltage (V_o). Also record the phase relationship between the input and output waveforms.

V_{in} = _____ p-p V_o = _____ p-p

Phase difference = _____ degrees

Question: What was the phase relationship between the amplifier input and output waveshapes?
Explain.

Step 3.   Based on the ac input and output voltages measured in Step 2, determine the overall voltage gain (A_V) of the amplifier.

Step 4.   Using the value for transistor emitter resistance (r_e) determined in Step 12, Experiment 11, calculate the expected voltage gain (A_V) for the amplifier in Figure 17-1.

Question: How did the measured amplifier voltage gain compare with the calculated gain?

Step 5.   Add a 1 kΩ resistor between node V_{in} and capacitor C_1 to simulate a 1 kΩ source resistance. Run the simulation and record the ac peak-to-peak input voltage (V_{in}) and the ac peak-to-peak output voltage (V_o). Adjust the oscilloscope settings as needed.

$V_{in} = $ _____ p-p $V_o = $ _____ p-p

Step 6.   Calculate the new voltage gain (A_V) based on the voltage readings in Step 5.

Question: How did adding 1 kΩ to the source resistance affect the overall voltage gain? **Explain.**

Step 7.   Move the Channel A oscilloscope lead to the transistor base. Run the simulation again
                 and record the ac peak-to-peak voltage on the transistor base (V_b). Adjust the
                 oscilloscope settings as needed.

                          $V_b =$ _____ p-p

Question: What was the relationship between the ac voltage at the source (V_{in}) and the ac voltage on
the base (V_b) with the 1kΩ resistor in series with the input source? **Explain.**

Step 8.   Based on the values of V_{in} and V_b, calculate the amplifier ac peak-to-peak input current
                 (I_{in}). Based on the value of V_b and the input current (I_{in}), calculate the amplifier ac input
                 resistance (R_{in}).

Step 9.   Based on the value of the transistor ac current gain (β_{ac}) measured in Step 10, Experiment
                 11, and the value of the transistor emitter resistance (r_e) measured in Step 12, Experiment
                 11, calculate the transistor ac input resistance (r_i). From the value of r_i and the bias
                 resistors R_1 and R_2, calculate the amplifier ac input resistance (R_{in}). (The amplifier ac input
                 resistance includes the base bias resistors in parallel with the transistor input resistance.)

Question: How did the measured value for amplifier input resistance (R_{in}) compare with the
calculated value?

Step 10.   Based on the value of V_b in Step 7 and V_o in Step 5, determine the voltage gain between the transistor base and the output.

Question: How did the voltage gain between the amplifier output and the transistor base compare with the overall voltage gain calculated in Step 3? **Explain.**

Step 11.   Replace the 1 kΩ resistor between node V_{in} and capacitor C_1 with a short. Replace the Channel A oscilloscope lead to node V_{in} to restore the circuit to that shown in Figure 17-1. Change the load resistor (R_L) to 2 kΩ, then run the simulation. Record the ac peak-to-peak input voltage (V_{in}) and the ac peak-to-peak output voltage (V_o). Adjust the oscilloscope settings as needed.

$V_{in} = $ _____ p-p $V_o = $ _____ p-p

Step 12.   Based on the voltage readings in Step 11, determine the new voltage gain.

Question: What effect did lowering the load resistance have on the overall gain of the amplifier? **Explain.**

Step 13.   Based on the difference between the output voltage in Step 2 and the output voltage in Step 11, calculate the output resistance (R_o) of the amplifier.

Question: What was the relationship between the amplifier output resistance (R_O) and the value of the collector resistor (R_C)?

Step 14.   Change R_L back to 200 kΩ and change the generator amplitude to 10 mV. On the oscilloscope change the Channel A input to 5 mV/Div, AC, and the Channel B input to 5 V/Div, DC. Change B to A (click A) in the oscilloscope Trigger section.

Step 15.   Run the simulation. Record the dc offset voltage of the output (blue) at node V_O. Move the Channel B oscilloscope wire to node V_C. Run the simulation again. Record the dc offset voltage of the output at node V_C.

DC offset at node V_O = _____

DC offset at node V_C = _____

Question: What effect did the output coupling capacitor have on the dc voltage offset of the ac output? **Explain.**

Step 16.   Return the Channel B oscilloscope wire to node V_O. Set the Channel B input on the oscilloscope to 20 mV/Div, AC. Remove the emitter bypass capacitor (C_3). Run the simulation and record the ac peak-to-peak input voltage (V_{in}) and the ac peak-to-peak output voltage (V_o).

V_{in} = _____ p-p V_o = _____ p-p

Step 17.   Based on the voltage readings in Step 16, determine the new voltage gain (A_V) with unbypassed emitter resistance.

Step18.   Based on the value of the transistor emitter resistance (r_e) determined in Step 12, Experiment 11 and the value of the unbypassed emitter resistance (R_E), calculate the expected voltage gain (A_V) of the amplifier with the bypass capacitor removed.

Questions: What effect did removing the emitter bypass capacitor have on the overall amplifier voltage gain? **Explain.**

How did the measured voltage gain compare with the calculated gain with the emitter bypass capacitor removed?

Step 19.   Double click transistor Q1, and then click *Edit Model*. Change the forward current gain coefficient (Bf) from 800 to 20, and then click *Change Part Model*. Click *OK* on the Models window to return to the circuit. You have changed the current gain (β) for the 2N3904 transistor. This will allow you to examine the effect that changing transistors would have on the voltage gain. Run the simulation and record the ac peak-to-peak input voltage (V_{in}) and the ac peak-to-peak output voltage (V_o).

$V_{in} = $ _____ p-p $V_o = $ _____ p-p

NOTE: If you are using this manual in a lab environment, you should change the 2N3904 transistor in place of changing the current gain (β).

Step 20.   Based on the voltage readings in Step 19, determine the voltage gain (A_V) with unbypassed emitter resistance.

Step 21.   Reconnect the bypass capacitor (C_3). The circuit should be as shown in Figure 17-1. Change the oscilloscope Channel B to 1 V/Div. Run the simulation and record the ac peak-to-peak input voltage (V_{in}) and the ac peak-to-peak output voltage (V_o).

$V_{in} = $ _____ p-p $V_o = $ _____ p-p

Step 22.   Based on the voltage readings in Step 21, determine the voltage gain (A_V) with bypassed emitter resistance.

Questions: How did the voltage gain in Step 20 (low β) compare with the voltage gain in Step 17 (high β) without the emitter bypass capacitor connected?

How did the voltage gain in Step 22 (low β) compare with the voltage gain in Step 3 (high β) with the emitter bypass capacitor connected?

What conclusion can you draw about gain stability without the emitter bypass capacitor compared to gain stability with the emitter bypass capacitor?

Step 23.   Double-click transistor Q1 and return the value of forward current gain coefficient (Bf) back to 800 for the 2N3904 transistor. Don't forget to click *Edit Model*, *Change Part Model*, and *OK*.

NOTE: If you are using this manual in a lab environment, you should change back to the original 2N3904 transistor.

Troubleshooting Problems

1.   Open Circuit file FIG17-2 and run the simulation. Locate the defective component and state the defect (short or open). You can use any instrument available and make any measurement desired.

            Defective component: _____   Defect: _____

2.   Open circuit file FIG17-3 and run the simulation. Locate the defective component and state the defect (short or open). You can use any instrument available and make any measurement desired.

            Defective component: _____   Defect: _____

3.   Open circuit file FIG17-4 and run the simulation. Locate the defective component and state the defect (short or open). You can use any instrument available and make any measurement desired.

            Defective component: _____   Defect: _____

4.   Open circuit file FIG17-5 and run the simulation. Locate the defective component and state the defect (short or open). You can use any instrument available and make any measurement desired.

          Defective component: _____   Defect: _____

5.   Open circuit file FIG17-6 and run the simulation. Locate the defective component and state the defect (short or open). You can use any instrument available and make any measurement desired.

          Defective component: _____   Defect: _____

6.   Open circuit file FIG17-7 and run the simulation. Locate the defective component and state the defect (short or open). You can use any instrument available and make any measurement desired.

          Defective component: _____   Defect: _____

18

Cascaded Common-Emitter Amplifier

Objectives:

1. Measure the dc base and collector voltages for each stage of a two-stage common-emitter (CE) amplifier and compare the measured voltages with the calculated values.
2. Determine the location of the dc operating point (Q-point) on the dc load line for each stage of a two-stage CE amplifier.
3. Measure the voltage gain of each stage of a two-stage CE amplifier and compare the measured voltage gain with the calculated value.
4. Measure the overall voltage gain of a two-stage CE amplifier and compare the measured voltage gain with the calculated value.
5. Observe the phase difference between the input and output waveshapes for a two-stage CE amplifier.

Materials:

Two 2N3904 bipolar junction transistors
One 10 V dc power supply
Capacitors: three 10 μF, two 47 μF
One dual-trace oscilloscope
One function generator
One digital multimeter
Resistors: two 250 Ω, two 750 Ω, three 5 kΩ, two 10 kΩ, two 50 kΩ, one 100 kΩ

Theory:

You should complete Experiments 11, 13, and 17 before attempting this experiment.

Several common-emitter amplifiers can be cascaded with the output of one stage driving the input of the next stage. Each stage inverts the input by 180 degrees. Therefore, an even number of stages will have an output that is in-phase with the input and an odd number of stages will have an output that is 180 degrees out-of-phase with the input. The purpose of cascading single-stage amplifiers into a **multistage amplifier** is to increase the overall gain. The **overall voltage gain (A_{VT})** of a multistage amplifier is the product of the individual gains. Therefore, for the two-stage common-emitter amplifier in Figure 18-1,

$$A_{VT} = A_{V1}A_{V2}$$

Based on the ac peak-to-peak output and input voltages, the **voltage gain** of an amplifier is determined by dividing the ac peak-to-peak output voltage by the ac peak-to-peak input voltage. Therefore, for the two-stage amplifier in Figure 18-1,

$$A_{V1} = \frac{V_{c1}}{V_{in}}$$

$$A_{V2} = \frac{V_o}{V_{c1}}$$

$$A_{VT} = \frac{V_o}{V_{in}}$$

where voltages V_{c1}, V_{in}, and V_o are the peak-to-peak ac voltages.

For each of the voltage-divider biased *npn* common-emitter amplifier configurations in Figure 18-1, the **dc load line** crosses the horizontal axis on the collector characteristic curve plot at a value of collector-emitter voltage (V_{CE}) equal to V_{cc}. It crosses the vertical axis at a value of collector current (I_C) equal to $V_{CC}/(R_C + R_e + R_E)$.

The **operating point (Q-point)** is located on the dc load line based on the value of the dc collector current (I_C) or the value of the dc collector-emitter voltage (V_{CE}).

The **dc base voltage (V_B)** for a voltage-divider biased *npn* common-emitter amplifier configuration is calculated using the voltage divider rule. If $\beta(R_{e1} + R_{E1}) \gg R_1$ in amplifier stage 1 in Figure 18-1, then

$$V_{B1} \cong \frac{V_{CC}R_1}{R_1 + R_2}$$

The **dc emitter current (I_E)** for a voltage-divider biased common-emitter amplifier configuration is calculated by first finding the dc emitter voltage (V_E). The emitter voltage is found by subtracting the base-emitter voltage (V_{BE}) from the base voltage (V_B). Then the dc emitter current is calculated by dividing the emitter voltage by the dc emitter resistance. Therefore, for amplifier stage 1 in Figure 18-1,

$$I_{E1} = \frac{V_{E1}}{R_{e1} + R_{E1}} = \frac{V_{B1} - V_{BE}}{R_{e1} + R_{E1}}$$

The **dc collector current (I_C)** can be estimated from the dc emitter current (I_E) as follows:

$$I_{C1} = I_{E1} - I_{B1} \cong I_{E1}$$

The **dc collector voltage (V_C)** is calculated by subtracting the dc voltage across the collector resistor (R_C) from the dc supply voltage (V_{CC}). Therefore,

$$V_{C1} = V_{CC} - I_{C1}R_{C1}$$

Because the biasing networks are the same for both amplifier stages,

$$I_{E2} = I_{E1}, \; I_{C2} = I_{C1}, \text{ and } V_{C2} = V_{C1}$$

The **expected voltage gain** of a one-stage common-emitter amplifier is calculated by dividing the ac collector resistance (R_c) by the total unbypassed ac emitter resistance. The ac collector resistance is equal to the parallel equivalent of the collector resistor (R_C) and the load resistance. In a cascaded multistage amplifier, the input resistance of each stage acts as a load resistance on the previous stage. Therefore, for amplifier stage 1 in Figure 18-1,

$$R_{c1} = \frac{R_{C1}R_{in2}}{R_{C1} + R_{in2}}$$

where the load resistance is equal to R_{in2}, which can be found from the equation

$$\frac{1}{R_{in2}} = \frac{1}{R_3} + \frac{1}{R_4} + \frac{1}{r_{i2}}$$

where $r_{i2} = \beta_2(r_{e2} + R_{e2})$, $r_{e2} \cong 25 \text{ mV}/I_{E2}(\text{mA})$, and R_{e2} is the unbypassed emitter resistance (stage 2).

The voltage gain of amplifier stage 1 (A_{V1}) is

$$A_{V1} = \frac{R_{c1}}{r_{e1} + R_{e1}}$$

where $r_{e1} \cong 25 \text{ mV}/I_{E1}(\text{mA})$ and R_{e1} is the unbypassed emitter resistance (stage 1).

The voltage gain of amplifier stage 2 (A_{V2}) is

$$A_{V2} = \frac{R_{c2}}{r_{e2} + R_{e2}}$$

where $R_{c2} = (R_{C2})(R_L)/(R_{C2} + R_L)$.

Figure 18-1 Two-Stage Common-Emitter Amplifier

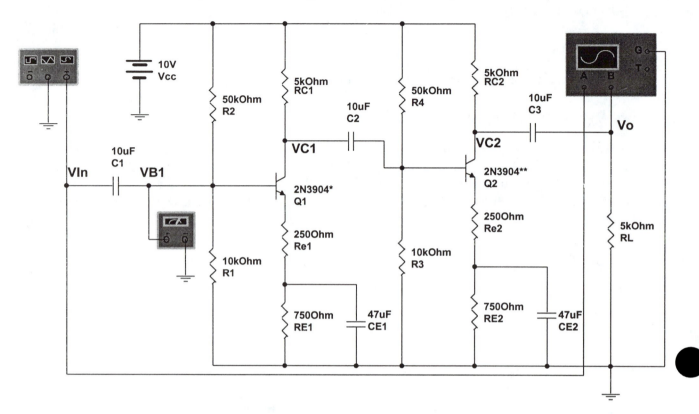

Procedure:

Step 1.   Open circuit file FIG18-1. Bring down the oscilloscope enlargement and make sure that the following settings are selected: Time base (Scale = 200 µs/Div, Xpos = 0, Y/T), Ch A (Scale = 10 mV/Div, Ypos = 0, AC), Ch B (Scale = 500 mV/Div, Ypos = 0, AC), Trigger (Pos edge, Level = 1 µV, Nor, A). Bring down the function generator enlargement and make sure that the following settings are selected: *Sine Wave*, Freq = 1 kHz, Ampl = 10 mV, Offset = 0. Bring down the multimeter enlargement and make sure that the following settings are selected: V, DC (——).   After the settings have been verified, click the box in the upper right corner of each instrument to return it to normal size.

Step 2.   Run the simulation. Double click the digital multimeter to bring down the enlargement. Record the dc voltage measured at node V_{B1}. Move the multimeter positive wire from node V_{B1} to node V_{C1}, then run the simulation and record the voltage reading. Next, move the multimeter wire back to node V_{B1}. (The biasing network for amplifier stage 2 is the same as for amplifier stage 1).

V_{B1} = _____ V_{C1} = _____

Step 3.   Based on the values of R_1, R_2, and V_{CC}, calculate the expected value of the dc voltage at node V_{B1}.

Question: How did the measured dc voltage on node V_{B1} in Step 2 compare with the calculated voltage in Step 3?

Step 4.   Based on the value of V_{B1} calculated in Step 3, estimate the dc voltage on the emitter (V_{E1}) of transistor Q_1. From this value, calculate I_{E1} and I_{C1} for transistor Q_1. Based on the values of I_{C1} and R_{C1}, calculate the expected dc voltage at node V_{C1}. (Assume $V_{BE} = 0.7$ V).

Question: How did the measured dc voltage on node V_{C1} in Step 2 compare with the calculated voltage in Step 4?

Step 5.   Draw the dc load line on a copy of the collector characteristic curve plotted in Experiment 11, Step 7. Locate the operating point (Q-point) on the dc load line.

Question: Was the operating point (Q-point) in the middle of the load line?

Step 6.   Bring down the oscilloscope enlargement and run the simulation again. Pause the simulation after steady state is reached. Record the ac peak-to-peak input voltage (V_{in}) and the ac peak-to-peak output voltage (V_o). Also record the phase difference between the input and output waveshapes. Move the Channel B oscilloscope lead to node V_{c1}, run the simulation until steady state is reached, then pause the simulation and record the ac peak-to-peak collector voltage (V_{c1}). Change the oscilloscope settings as needed.

$V_{in} = $ _____ $V_o = $ _____ $V_{c1} = $ _____

Phase difference = _____ degrees

Question: What was the phase difference between the input and output waveshapes for the two-stage amplifier? **Explain.**

Step 7.   Based on the readings in Step 6, determine the voltage gain of amplifier stage 1 (A_{V1}) and the voltage gain of amplifier stage 2 (A_{V2}).

Step 8.   Based on the readings in Step 6, determine the overall voltage gain (A_{VT}) of the two-stage amplifier.

Step 9.   Calculate the overall voltage gain (A_{VT}) of the two-stage amplifier from A_{V1} and A_{V2}.

Question: How did the measured value of the overall amplifier voltage gain in Step 8 compare with the value calculated from the individual amplifier stages in Step 9?

Step 10.   From the values of R_{C1}, R_{e1}, r_{e1}, r_{e2}, R_3, R_4, and r_{i2}, calculate the expected voltage gain of amplifier stage 1 (A_{V1}). (Assume $\beta_2 = 200$).

Question: How did the measured voltage gain of amplifier stage 1 in Step 7 compare with the calculated value in Step 10?

Step 11.   From the values of R_{C2}, R_L, r_{e2}, and R_{e2}, calculate the expected voltage gain of amplifier stage 2 (A_{V2}).

Question: How did the measured voltage gain of amplifier stage 2 in Step 7 compare with the calculated value in Step 11?

Step 12.   Change the value of R_L to 100kΩ. Run the simulation until steady state is reached, then pause the simulation. Record the ac peak-to-peak input voltage (V_{in}) and the ac peak-to-peak collector voltage (V_{c1}). Move the oscilloscope Channel B lead to the node labeled V_o, run the simulation until steady state is reached, then pause the simulation. Record the ac peak-to-peak output voltage (V_o). Adjust the oscilloscope settings as needed.

V_{in} = _____ V_{c1} = _____ V_o = _____

Step 13.   Based on the readings in Step 12, determine the overall voltage gain (A_{VT}) of the two-stage amplifier.

Step 14.   Based on the readings in Step 12, determine the voltage gain of amplifier stage 1 (A_{V1}) and amplifier stage 2 (A_{V2}).

Questions: When the value of the load resistor (R_L) was increased in Step 12, what effect did it have on the overall voltage gain of the two-stage amplifier?   The voltage gain of amplifier stage 1?   The voltage gain of amplifier stage 2? **Explain.**

Troubleshooting Problems

1.   Open circuit file FIG18-2 and run the simulation. Locate the defective component and state the defect (short or open). You can use any instrument available and make any measurement desired.

           Defective component: _____   Defect: _____

2.   Open circuit file FIG18-3 and run the simulation. Locate the defective component and state the defect (short or open). You can use any instrument available and make any measurement desired.

           Defective component: _____   Defect: _____

3.   Open circuit file FIG18-4 and run the simulation. Locate the defective component and state the defect (short or open). You can use any instrument available and make any measurement desired.

Defective component: _____   Defect: _____

4.   Open circuit file FIG18-5 and run the simulation. Locate the defective component and state the defect (short or open). You can use any instrument available and make any measurement desired.

Defective component: _____   Defect: _____

5.   Open circuit file FIG18-6 and run the simulation. Locate the defective component and state the defect (short or open). You can use any instrument available and make any measurement desired.

Defective component: _____   Defect: _____

19

Name _____

Date _____

Emitter-Follower

Objectives:

1. Measure the dc base and emitter voltages for a common-collector (emitter-follower) circuit and compare the measured voltages with the calculated values.
2. Determine the location of the dc operating point (Q-point) on the dc load line.
3. Measure the voltage gain of a common-collector (emitter-follower) circuit and compare the measured voltage gain with the calculated value.
4. Measure the ac input resistance of a common-collector (emitter-follower) circuit and compare the measured and calculated values.
5. Measure the ac output resistance of a common-collector (emitter-follower) circuit.
6. Observe the phase difference between the input and output waveshapes for a common-collector (emitter-follower) circuit.

Materials:

One 2N3904 bipolar junction transistor
One 10 V dc power supply
Capacitors: one 1 µF, one 100 µF
One dual-trace oscilloscope
One function generator
One digital multimeter
Resistors: one 500 Ω, one 2 kΩ, one 10 kΩ, two 20 kΩ, one 50 kΩ

Theory:

In a **common-collector (CC) amplifier** configuration, as shown in Figure 19-1, the transistor collector is at ac ground potential through the power supply filter capacitor, the ac input is applied to the transistor base through a coupling capacitor (C_1), and the ac load resistance (R_L) is connected to the emitter through a coupling capacitor (C_2). The **input coupling capacitor (C_1)** prevents the input circuit from affecting the dc base bias voltage, and the **output coupling capacitor (C_2)** prevents the load resistance (R_L) from affecting the dc emitter bias voltage. When the input signal voltage increases, both the base current and the emitter current increase. The increasing emitter current causes the emitter voltage (output) to increase. Therefore, the emitter (output) voltage follows (is in-phase with) the input voltage in a CC amplifier. This is the reason that a CC amplifier is normally referred to as an **emitter-follower**. The **voltage gain** of an emitter-follower is normally slightly less than one (1). Its main advantages are a **high input resistance** and a **low output resistance**. This makes it useful as a buffer to minimize the loading effects on a circuit that is driving a low resistance load.

For the emitter-follower circuit in Figure 19-1, the **dc load line** crosses the horizontal axis on the collector characteristic curve plot at a value of collector-emitter voltage (V_{CE}) equal to V_{CC}. It crosses the vertical axis at a value of collector current (I_C) equal to V_{CC}/R_E.

The **operating point (Q-point)** is located on the dc load line based on the value of the collector current (I_C) or the value of the collector-emitter voltage (V_{CE}).

The **dc base voltage (V_B)** for the emitter-follower circuit in Figure 19-1 is calculated using the voltage divider rule. If $\beta(R_E) \gg R_1$, then

$$V_B \cong \frac{V_{CC}R_1}{R_1 + R_2}$$

The **dc emitter current (I_E)** is calculated by dividing the dc emitter voltage (V_E) by the emitter resistor value (R_E). Therefore,

$$I_E = \frac{V_E}{R_E}$$

The **dc collector current (I_C)** can be estimated from the dc emitter current (I_E) as follows:

$$I_C = I_E - I_B \cong I_E$$

As in all amplifiers, the **voltage gain** is determined by dividing the ac peak-to-peak output voltage (V_o) by the ac peak-to-peak input voltage (V_{in}). The **expected voltage gain (A_V)** of the one-stage emitter-follower in Figure 19-1 is calculated from the equation

$$A_V = \frac{R_e}{r_e + R_e}$$

where $r_e \cong 25 \text{ mV}/I_E(\text{mA})$ and R_e is the external ac emitter resistance and can be found from

$$R_e = \frac{R_E R_L}{R_E + R_L}$$

Notice that the voltage gain (A_V) is always less than one (1) because the denominator in the voltage gain equation will always be larger than the numerator. If the transistor ac emitter resistance (r_e) is much less than the external ac emitter reisistance (R_e), the voltage gain is close to one (1).

The **ac input resistance (R_{in})** of the emitter-follower in Figure 19-1 can be determined by placing a 10 kΩ resistor between V_{in} and capacitor C_1 and dividing the ac peak-to-peak input voltage on the base (V_b) by the ac peak-to-peak input current (I_{in}). Therefore,

$$R_{in} = \frac{V_b}{I_{in}}$$

where $I_{in} = (V_{in} - V_b)/10 \text{ k}\Omega$

The ac input resistance (R_{in}) of the emitter-follower in Figure 19-1 can be calculated from the circuit component values by first finding the **transistor input resistance (r_i)** and combining this resistance with the parallel equivalent resistance of the base biasing network. The transistor input resistance is calculated from

$$r_i = (\beta)(R_e + r_e)$$

The amplifier ac input resistance (R_{in}) is then found from the equation

$$\frac{1}{R_{in}} = \frac{1}{R_1} + \frac{1}{R_2} + \frac{1}{r_i}$$

where R_1 and R_2 are the base biasing network resistors.

Figure 19-1 Emitter-Follower

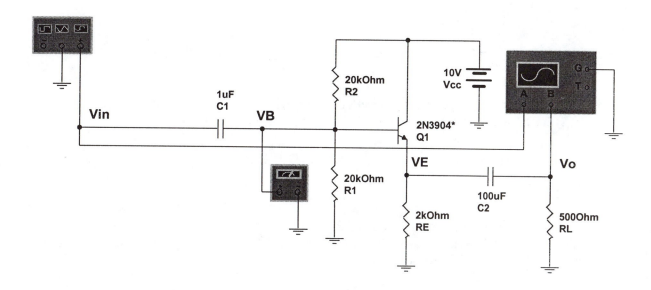

Procedure:

Step 1.   Open circuit file FIG19-1. Bring down the oscilloscope enlargement and make sure that the following settings are selected: Time base (Scale = 20 μs/Div, Xpos = 0, Y/T), Ch A (Scale = 200 mV/Div, Ypos = 0, AC), Ch B (Scale = 200 mV/Div, Ypos = 0, AC), Trigger (Pos edge, Level = 0, Sing, A). Bring down the function generator enlargement and make sure that the following settings are selected: *Sine Wave*, Freq = 10 kHz, Ampl = 500 mV, Offset = 0. Bring down the multimeter enlargement and make sure that the following settings are selected: V, DC (——). Run the simulation. Click *pause* to pause the simulation when steady state is reached on the multimeter. Record the dc voltage measured on the base (V_B). Move the multimeter positive wire from node V_B to node V_E. Run the simulation and record the dc voltage (V_E) on node V_E. Then move the multimeter wire back to node V_B.

$V_B =$ _____ $V_E =$ _____

Step 2.   Based on the values of R_1, R_2, and V_{CC}, calculate the expected value of the dc voltage on the base (V_B).

Question: How did the measured value of the dc voltage on the base (V_B) compare with the calculated value?

Step 3.   Based on the value of the dc emitter voltage (V_E), calculate the emitter current (I_E). From the value of the emitter current, estimate the collector current (I_C).

Step 4.   Draw the dc load line on a copy of the collector characteristic curve plotted in Experiment 11, Step 7. Based on the collector current (I_C) calculated in Step 3, locate the operating point (Q-point) on the dc load line.

Question: Was the operating point (Q-point) near the middle of the load line?

Step 5. Bring down the oscilloscope and run the simulation. Pause the simulation after one screen display on the oscilloscope. Record the ac peak-to-peak input (red) voltage (V_{in}) and the ac peak-to-peak output (blue) voltage (V_o). Also record the phase difference between the input and output waveshapes.

$V_{in} = $ _____ $V_o = $ _____

Phase difference = _____ degrees

Question: What was the phase difference between the input and output waveshapes for the emitter-follower? **Explain.**

Step 6. Based on the readings in Step 5, determine the voltage gain (A_V) of the emitter-follower.

Step 7. Based on the value of the amplifier ac emitter resistance (R_e) and the transistor emitter resistance (r_e), calculate the expected voltage gain (A_V) of the emitter-follower. (Remember that R_e is the parallel combination of R_E and R_L).

Questions: How did the measured voltage gain of the emitter-follower compare with the calculated value? Was the gain less than 1?

Step 8. Insert a 10 kΩ resistor between node V_{in} and capacitor C_1. Connect the Channel B oscilloscope lead to node V_B. Run the simulation. Click *pause* to pause the simulation after one screen display on the oscilloscope. Record the ac peak-to-peak input voltage (V_{in}) and the ac peak-to-peak base voltage (V_b).

$V_{in} = $ _____ $V_b = $ _____

Step 9.   Based on the readings in Step 8, determine the ac input current (I_{in}). Using this value and
            the value of V_b, determine the ac input resistance (R_{in}).

Step 10.   Calculate the transistor ac input resistance (r_i). (Assume that $\beta = 200$). Based on the
            value of R_1, R_2, and the transistor ac input resistance (r_i), calculate the expected ac input
            resistance (R_{in}) for the emitter-follower. Remember that the ac emitter resistance (R_e) is
            the parallel combination of R_E and R_L.

Questions: How did the measured emitter-follower ac input resistance (R_{in}) in Step 9 compare with
the calculated value in Step 10? Was the ac input resistance (R_{in}) large or small compared to the input
resistance of a common-emitter amplifier?

Step 11.   Move the Channel B oscilloscope lead back to node V_O and change the load resistor (R_L) to
            50 kΩ. Run the simulation. Click *pause* to pause the simulation after one screen display
            on the oscilloscope. Record the ac peak-to-peak output voltage (V_o). Change the Trigger
            selection to Nor on the oscilloscope. Next, keep reducing the value of the load resistor (R_L)
            and running the simulation until the ac peak-to-peak output voltage (V_o) is equal to one-
            half the value previously measured when R_L was 50 kΩ. Adjust the oscilloscope as needed
            and run the simulation until the oscilloscope curve plot reaches steady-state. Record the
            new value of the load resistor (R_L). This is equal to the emitter-follower ac output
            resistance (R_o). (Note: You may need to reduce the value of R_L below 100 Ω.)

            $V_o(R_L = 50\ k\Omega) = $ _____ $R_o = $ _____

Questions: Was the measured value of the ac output resistance (R_o) of the emitter-follower high or low compared to the output resistance of the common-emitter amplifier? Was this expected?

What is the main advantage of the emitter-follower in terms of the ac input resistance and the ac output resistance? What is the main purpose of an emitter-follower?

Troubleshooting Problems

1.   Open circuit file FIG19-2 and run the simulation. Locate the defective component and state the defect (short or open). You can use any instrument available and make any measurement desired.

           Defective component: _____   Defect: _____

2.   Open circuit file FIG19-3 and run the simulation. Locate the defective component and state the defect (short or open). You can use any instrument available and make any measurement desired.

           Defective component: _____   Defect: _____

3.   Open circuit file FIG19-4 and run the simulation. Locate the defective component and state the defect (short or open). You can use any instrument available and make any measurement desired.

           Defective component: _____   Defect: _____

4.   Open circuit file FIG19-5 and run the simulation. Locate the defective component and state the defect (short or open). You can use any instrument available and make any measurement desired.

           Defective component: _____   Defect: _____

EXPERIMENT

Name_____

Date_____

20

Large-Signal Class A Amplifier

Objectives:

1. Determine the dc load line and locate the operating point (Q-point) on the dc load line for a large-signal class A common-emitter amplifier.
2. Determine the ac load line for a large-signal class A common-emitter amplifier.
3. Center the operating point (Q-point) on the ac load line.
4. Determine the maximum ac peak-to-peak output voltage swing before peak clipping occurs and compare the measured value with the expected value.
5. Observe nonlinear distortion of the output waveshape.
6. Measure the large-signal voltage gain of a class A common-emitter amplifier and compare the measured and calculated values.
7. Measure the maximum undistorted output power for a class A amplifier.
8. Determine the amplifier efficiency of a class A amplifier.

Materials:

One 2N3904 bipolar junction transistor
One 20 V dc power supply
Capacitors: two 10 µF, one 470 µF
One digital multimeter
One function generator
One dual-trace oscilloscope
Resistors: one 5 Ω, one 95 Ω, two 100 Ω, one 1 kΩ, one 2.4 kΩ

Theory:

A **power amplifier** is a **large-signal amplifier** that normally provides power to an antenna or speaker in the final stage of a communications transmitter or receiver. When an amplifier is biased so that it operates in the linear region (between saturation and cutoff) of the transistor collector characteristic curve plot for the full 360 degrees of the input sine wave cycle, it is classified as a **class A amplifier**. This means that collector current flows during the full input sine wave cycle, making class A amplifiers the **least efficient** of the different classes of large-signal amplifiers. In a large-signal amplifier, the input signal causes the operating point (Q-point) to move over a much larger portion of the ac load line than in a small-signal amplifier. Therefore, large-signal class A amplifiers require the operating point to be as close as possible to the center of the ac load line to avoid clipping of the output waveform. In a class A amplifier, the output voltage waveform has the same shape as the input voltage waveform, making it the **most linear** of the different classes of large-signal amplifiers. Most small-signal amplifiers are class A amplifiers. In this experiment you will study a large-signal class A amplifier.

For the large-signal class A common-emitter amplifier shown in Figures 20-1 and 20-2, the **dc collector-emitter voltage (V_{CE})** can be calculated from

$$V_{CE} = V_C - V_E$$

The **dc collector current (I_C)** can be found by calculating the current in the collector resistor (R_C). Therefore,

$$I_C = \frac{V_{CC} - V_C}{R_C}$$

The **ac collector resistance (R_c)** is equal to the parallel equivalent of the collector resistor (R_C) and the load resistor (R_L). Therefore,

$$R_c = \frac{R_C R_L}{R_C + R_L}$$

The **ac load line** has a slope of $1/(R_c + R_e)$ and crosses the dc load line Q-point. (See Theory in Experiment 18 to locate the dc load line and the Q-point). The ac load line crosses the horizontal axis of the transistor collector characteristic curve plot at V_{CE} equal to $V_{CEQ} + (I_{CQ})(R_c + R_e)$, where V_{CEQ} is the collector-emitter voltage at the Q-point and I_{CQ} is the collector current at the Q-point.

The **amplifier voltage gain** is measured by dividing the ac peak-to-peak output voltage (V_o) by the ac peak-to-peak input voltage (V_{in}). The **expected amplifier voltage gain (A_V)** for a common-emitter amplifier is calculated from

$$A_V = \frac{R_c}{r_e + R_e}$$

where R_c is the ac collector resistance, $r_e \cong 25 \text{ mV}/I_E(\text{mA})$, and R_e is the unbypassed emitter resistance.

In order to center the Q-point on the ac load line, you must try different values of R_E until V_{CEQ} is equal to $(I_{CQ})(R_c + R_e)$, where $I_{CQ} \cong I_{EQ} = V_E/(R_E + R_e)$, $V_{CEQ} = V_{CC} - I_{CQ}(R_E + R_e + R_C)$, R_c is equal to the ac collector resistance, and R_C is equal to the dc collector resistance.

The **amplifier output power (P_o)** can be calculated as follows:

$$P_o = \frac{V_{rms}^2}{R_L} = \frac{V_{o(p-p)}^2}{8R_L}$$

where $V_{o(p-p)}$ is the peak-to-peak output voltage and $V_{rms} = V_{o(p-p)}/2\sqrt{2}$.

The **efficiency (η)** of a large-signal amplifier is equal to the maximum output power (P_o) divided by the power supplied by the source (P_S) times 100%. Therefore,

$$\eta = \frac{P_o}{P_S}(100\%)$$

where $P_S = (V_{CC})(I_S)$. The current at the source (I_S) is determined from

$$I_S = I_{12} + I_{CQ}$$

where $I_{12} = V_{CC}/(R_1 + R_2)$. *Note*: I_{12} is the current in resistors R_1 and R_2, neglecting the base current.

Figure 20-1 Large-Signal Class A Amplifier, DC Analysis

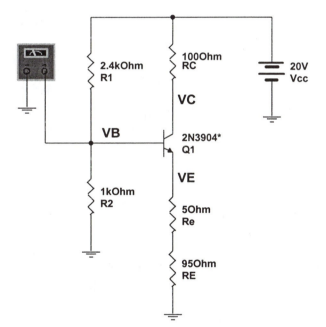

Figure 20-2 Large-Signal Class A Amplifier

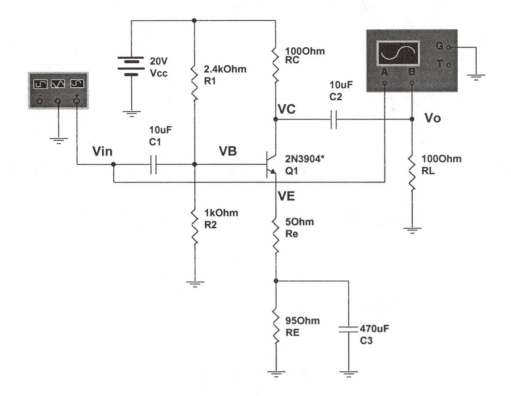

Procedure:

Step 1.   Open circuit file FIG20-1. Bring down the multimeter enlargement and make sure that the following settings are selected: V, DC (———). Run the simulation. After steady state has been reached, record the dc base voltage (V_B). Next , move the multimeter positive lead to node V_E, run the simulation, and record the dc emitter voltage (V_E). Then move the multimeter positive lead to node V_C, run the simulation, record the dc collector voltage (V_C), then pause the simulation.

$V_B =$ _____ $V_E =$ _____ $V_C =$ _____

Step 2.   Based on the voltages recorded in Step 1, calculate the dc collector-emitter voltage (V_{CE}) and the dc collector current (I_C).

Step 3.   Draw the dc load line on the graph provided. Based on the calculations in Step 2, locate
          the operating point (Q-point) on the dc load line.

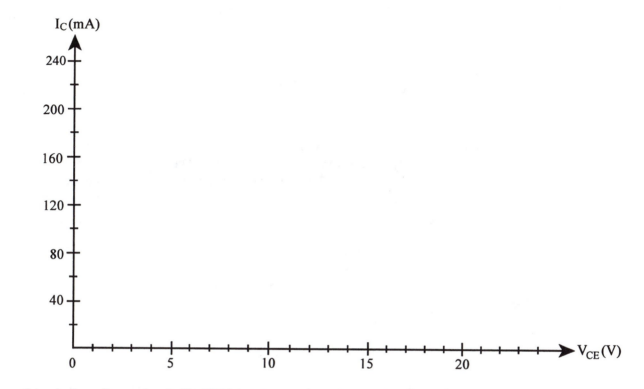

Step 4.   Open circuit file FIG20-2. Bring down the oscilloscope enlargement and make sure that the
          following settings are selected: Time base (Scale = 100 μs/Div, Xpos = 0, Y/T), Ch A (Scale
          = 200 mV/Div, Ypos = 0, AC), Ch B (Scale = 2 V/Div, Ypos = 0, AC), Trigger (Pos edge,
          Level = 0, Nor, A). Bring down the function generator enlargement and make sure that the
          following settings are selected: *Sine Wave*, Freq = 2 kHz, Ampl = 250 mV, Offset = 0.
          Based on the value of R_C and R_L, calculate the ac collector resistance (R_c), and then draw
          the ac load line through the Q-point on the graph in Step 3.

Questions: Is the operating point (Q-point) in the center of the dc load line? In the center of the ac
load line?

Why is it necessary for the Q-point to be in the center of the ac load line for large-signal inputs?

Step 5.   Run the simulation. Keep increasing the input signal voltage on the function generator until output (blue curve plot) peak distortion begins to occur. Then reduce the input signal level until there is no longer any output peak distortion. *Pause* the simulation and record the maximum undistorted ac peak-to-peak output voltage (V_o) and the ac peak-to-peak input voltage (V_{in}).

$V_{in} = $ _____ $V_o = $ _____

Step 6.   Based on the voltages measured in Step 5, determine the voltage gain (A_V) of the amplifier.

Step 7.   Calculate the expected voltage gain (A_V) based on the value of the ac collector resistance (R_c), the unbypassed emitter resistance (R_e), and the transistor ac emitter resistance (r_e), where $r_e \cong 25mV/I_E(mA)$.

Questions: How did the measured amplifier voltage gain compare with the calculated voltage gain?

What effect does unbypassed emitter resistance have on the amplifier voltage gain? On the voltage gain stability?

Step 8.   Calculate the value of R_E required to center the Q-point on the ac load line. *Hint*: Try different values of R_E until $V_{CEQ} = (I_{CQ})(R_c + R_e)$ at the new Q-point. See Theory for details.

Question: Did you need to increase or decrease R_E to center the Q-point on the ac load line? **Explain why.**

Step 9.   Change R_E to the value calculated in Step 8 and repeat the procedure in Step 5. Record the maximum undistorted ac peak-to-peak output (blue curve plot) voltage (V_o) and the ac peak-to-peak input voltage (V_{in}) for this centered Q-point.

$V_{in} =$ _____ $V_o =$ _____

Question: How did the maximum undistorted peak-to-peak output voltage measured in Step 9, for the centered Q-point, compare with the maximum undistorted peak-to-peak voltage measured in Step 5, for the original Q-point that was not centered?

Step 10.   Calculate the new dc values for I_{CQ} and V_{CEQ} for the new value of R_E. Locate the new dc load line and the new Q-point. Draw the new ac load line through the new Q-point.

Question: Was the new Q-point near the center of the new ac load line?

Step 11.   Based on the new centered Q-point on the new ac load line, estimate what the maximum ac peak-to-peak output voltage (V_o) should be before output clipping occurs.

$$V_o = \underline{\hspace{2cm}}$$

Question: How did the maximum undistorted peak-to-peak output voltage measured in Step 9, for the centered Q-point, compare with the expected maximum estimated in Step 11? **Explain any difference.**

Step 12.   Based on the maximum undistorted ac peak-to-peak output voltage (V_o) measured in Step 9, calculate the maximum undistorted output power (P_o) to the load (R_L).

Step 13.   Based on the supply voltage (V_{CC}), the new collector current at the new operating point (I_{CQ}), and the bias resistor current (I_{12}), calculate the power supplied by the dc voltage source (P_S).

Step 14.   Based on the power supplied by the dc voltage source (P_S) and the maximum undistorted output power (P_o) calculated in Step 12, calculate the efficiency (η) of the amplifier.

Question: Is the efficiency of a class A amplifier high or low? **Explain.**

Troubleshooting Problems

1.　Open circuit file FIG20-3 and run the simulation. Locate the defective component and state the defect (short or open). You can use any instrument available and make any measurement desired.

　　　　Defective component: _____　　　Defect: _____

2.　Open the circuit file FIG20-4 and run the simulation. Locate the defective component and state the defect (short or open). You can use any instrument available and make any measurement desired.

　　　　Defective component: _____　　　Defect: _____

3.　Open circuit file FIG20-5 and run the simulation. Locate the defective component and state the defect (short or open). You can use any instrument available and make any measurement desired.

　　　　Defective component: _____　　　Defect: _____

4.　Open circuit file FIG20-6 and run the simulation. Locate the defective component and state the defect (short or open). You can use any instrument available and make any measurement desired.

　　　　Defective component: _____　　　Defect: _____

5.　Open circuit file FIG20-7 and run the simulation. Locate the defective component and state the defect (short or open). You can use any instrument available and make any measurement desired.

　　　　Defective component: _____　　　Defect: _____

Name_____

Date_____

21

Class B Push-Pull Amplifier

Objectives:

1. Determine the dc load line and locate the operating point (Q-point) on the dc load line for a class B push-pull amplifier.
2. Determine the ac load line for a class B push-pull amplifier.
3. Observe crossover distortion of the output waveshape and learn how to eliminate it.
4. Determine the maximum ac peak-to-peak output voltage swing before peak clipping occurs for a class B push-pull amplifier and compare the measured value with the expected value.
5. Compare the maximum undistorted ac peak-to-peak output voltage swing for a class B amplifier with the maximum for a class A amplifier.
6. Measure the large-signal voltage gain of a class B push-pull amplifier.
7. Measure the maximum undistorted output power for a class B push-pull amplifier.
8. Determine the amplifier efficiency of a class B push-pull amplifier.

Materials:

One 2N3904 *npn* bipolar junction transistor
One 2N3906 *pnp* bipolar junction transistor
Two 1N4001 diodes
One 20 V dc power supply
Capacitors: two 10 μF, one 100 μF
One digital multimeter
One function generator
One dual-trace oscilloscope
One 0–1 mA milliammeter
Resistors: one 100 Ω, two 5 kΩ

Theory:

When an amplifier is biased at **cutoff** so that it operates in the linear region of the collector characteristic curves for one-half cycle of the input sine wave (180°) and is at cutoff for the other half of the input cycle (180°), it is classified as a **class B amplifier**. In order to produce a complete reproduction of the input waveshape, a matched complementary pair of transistors in a **push-pull** configuration, as shown in Figure 21-1, is necessary. In a **class B push-pull amplifier**, each transistor conducts during opposite halves of the input cycle. When the input is zero, both transistors are at cutoff ($I_C = 0$). This makes a class B amplifier much **more efficient** than a class A amplifier, in which the transistor conducts for the entire input cycle (360°). The main **disadvantage** of a class B amplifier is that it is not as linear as a class A amplifier, producing a more distorted output.

In the **class B push-pull emitter-follower** configuration in Figure 21-1, both transistors are biased at cutoff. When a transistor is biased at cutoff, the input signal must exceed the base-emitter junction potential (V_{BE}) before it can conduct. Therefore, in the push-pull configuration in Figure 21-1 there is a time interval during the input transition from positive to negative or negative to positive when the transistors are not conducting, resulting in what is known as **crossover distortion**. The dc biasing network in Figure 21-2 will reduce the crossover distortion by biasing the transistors slightly above cutoff. Also, when the characteristics of the diodes (D_1 and D_2) are matched to the transistor characteristics, a stable dc bias is maintained over a wide temperature range.

The **dc load line** for each transistor in Figure 21-2 is a **vertical line** crossing the horizontal axis at $V_{CE} = V_{CC}/2$. The load line is vertical because there is no dc resistance in the collector or emitter circuit (slope of the dc load line is the inverse of the dc collector and emitter resistance). The **Q-point** on the dc load line for each transistor is close to cutoff ($I_C = 0$). The **dc collector-emitter voltages** for the two transistors in Figure 21-2 can be determined from the value of V_E using the equations

$$V_{CE1} = V_{CC} - V_E$$

and

$$V_{CE2} = V_E - 0 = V_E$$

The complete **class B push-pull amplifier** is shown in Figure 21-3. Capacitors C_1, C_2, and C_3 are **coupling capacitors** to prevent the transistor dc bias voltages from being affected by the input circuit or the load circuit. The **ac load line** for each transistor should have a slope of $1/R_L$ (the ac equivalent resistance in the emitter circuit is R_L), cross the horizontal axis at $V_{CC}/2$, and cross the vertical axis at $I_{C(sat)} = V_{CC}/2R_L$. The **Q-point** on the ac load line should be close to cutoff ($I_C = 0$) for each transistor. When one of the transistors is conducting, its operating point (Q-point) moves up the ac load line. The voltage swing of the conducting transistor can go all the way from cutoff to saturation. On the alternate half cycle the other transistor can also swing from cutoff to saturation. Therefore, the theoretical **maximum peak-to-peak output voltage** is equal to $2(V_{CC}/2) = V_{CC}$. The **amplifier voltage gain** is measured by dividing the ac peak-to-peak output voltage (V_o) by the ac peak-to-peak input voltage (V_{in}). Because the push-pull amplifier in Figure 21-3 is an **emitter-follower** configuration, the voltage gain should be close to **unity (1)**. This is no problem for large-signal amplifiers because they are used mostly for **power amplification** rather than voltage amplification.

The **amplifier output power (P_o)** can be calculated as follows:

$$P_o = \frac{V_{rms}^2}{R_L} = \frac{V_{o(p-p)}^2}{8R_L}$$

where $V_{o(p-p)}$ is the peak-to-peak output voltage and $V_{rms} = V_{o(p-p)}/2\sqrt{2}$.

The **percent efficiency (η)** of a power amplifier is equal to the maximum output power (P_o) divided by the dc power supplied by the source (P_{DC}) times 100%. Therefore,

$$\eta = \frac{P_o}{P_{DC}}(100\%)$$

where $P_{DC} = (V_{CC})(I_{CC})$. The power supply current (I_{CC}) is determined from

$$I_{CC} = I_{C(AVG)} = \frac{I_{C(sat)}}{\pi}$$

where $I_{C(sat)} = V_{CC}/2R_L$ and $I_{C(AVG)}$ is the average value of the half wave collector current.

Note: I_{RB1} is normally much less than $I_{C(AVG)}$ and can be neglected.

Figure 21-1 Class B Push-Pull Amplifier with Crossover Distortion

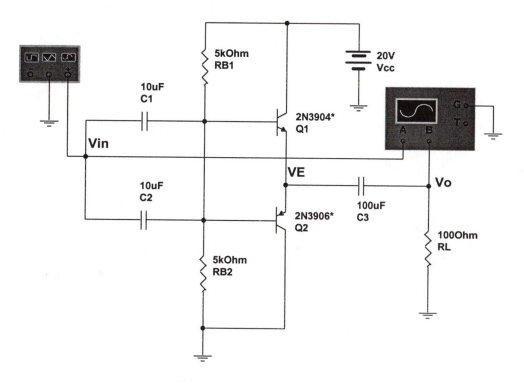

Figure 21-2 Class B Push-Pull Amplifier, DC Analysis

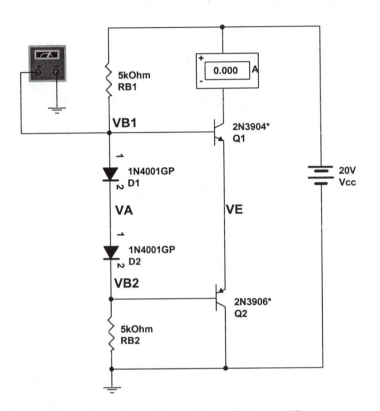

Figure 21-3 Class B Push-Pull Amplifier

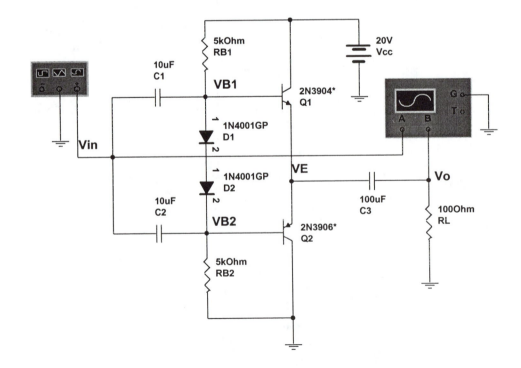

Procedure:

Step 1.   Open circuit file FIG21-1. Bring down the oscilloscope enlargement and make sure that the following settings are selected: Time base (Scale = 200 µs/Div, Xpos = 0, Y/T), Ch A (Scale = 2 V/Div, Ypos = 0, AC), Ch B (Scale = 2 V/Div, Ypos = 0, AC), Trigger (Pos edge, Level = 0, Nor, A). Bring down the function generator enlargement and make sure that the following settings are selected: *Sine Wave*, Freq = 1 kHz, Ampl = 4 V, Offset = 0. Run the simulation to ten full screen displays on the oscilloscope, then *pause* the simulation. You are plotting the amplifier input (red) and output (blue) on the oscilloscope. Notice the crossover distortion of the output waveshape (blue curve). Draw this waveshape in the space provided and **note the crossover distortion**.

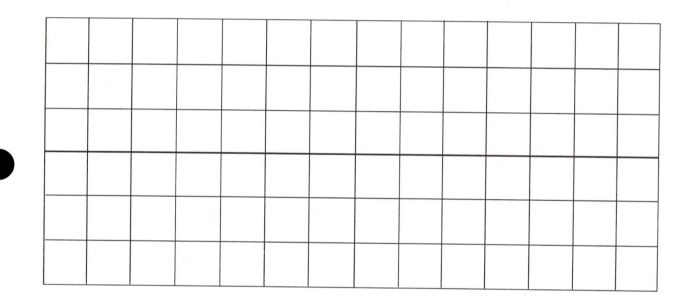

Step 2.   Open circuit file FIG21-2. Bring down the multimeter enlargement and make sure that the following settings are selected: V, DC (——). Run the simulation. After steady state has been reached on the multimeter, record the dc base 1 voltage (V_{B1}). Then move the multimeter positive lead to node V_{B2}, then node V_E, and then node V_A and run the simulation for each reading and record those dc voltages. Also record the dc collector current (I_C), then pause the simulation.

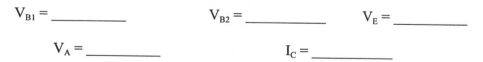

$V_{B1} =$ _____ $V_{B2} =$ _____ $V_E =$ _____

$V_A =$ _____ $I_C =$ _____

Step 3.   Based on the voltages recorded in Step 2, calculate the dc collector-emitter voltage (V_{CE}) for each transistor.

Step 4.   Draw the dc load line on the graph and locate the operating point (Q-point) on the dc load
              line based on the data in Step 2 and the calculations in Step 3.

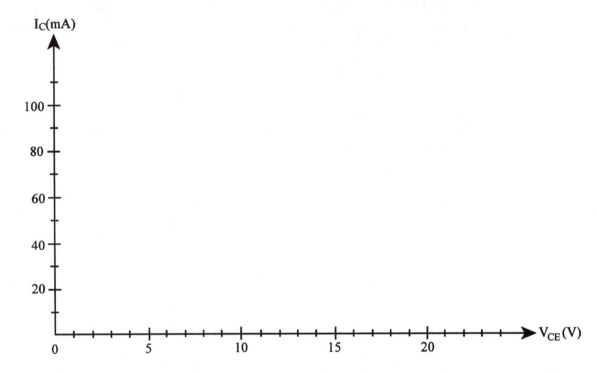

Step 5.   Open circuit file FIG21-3. Make sure the instrument settings are the same as Step 1.
              Based on the values of V_{CC} and R_L, draw the ac load line on the graph in Step 4.

Questions: Where was the operating point (Q-point) on the dc load line? On the ac load line?
Explain.

What is the relationship between the dc load line and the ac load line? **Explain.**

Step 6.   Run the simulation. Notice that there is very little crossover distortion of the output
             waveshape (blue). Keep increasing the input signal voltage until output peak distortion
             occurs. Then reduce the input signal level until there is no longer any distortion. Pause
             the simulation and record the maximum undistorted ac peak-to-peak output voltage (V_o)
             and the ac peak-to-peak input voltage (V_{in}). Adjust the oscilloscope settings as needed.

             $V_{in} =$ _____ $V_o =$ _____

Questions: What caused the crossover distortion in Step 1? What does the addition of diodes D1 and
D2 accomplish?

How did the maximum undistorted peak-to-peak output voltage for the class B amplifier, measured in
Step 6, compare with the maximum undistorted peak-to-peak output voltage for the class A amplifier,
measured in Experiment 20, Step 9?

Step 7.   Based on the voltages measured in Step 6, calculate the voltage gain of the amplifier.

Question: How did the measured amplifier voltage gain compare with the expected value for an
emitter-follower circuit?

Step 8.   Based on the ac load line and Q-point located on the graph in Step 4, estimate what the maximum ac peak-to-peak output voltage (V_o) should be before output clipping occurs. Record your answer.

$$V_o = \underline{\hspace{2cm}} \text{ p-p}$$

Question: How did the maximum undistorted peak-to-peak output voltage measured in Step 6 compare with the expected maximum estimated in Step 8?

Step 9.   Based on the maximum undistorted ac peak-to-peak output voltage measured in Step 6, calculate the maximum undistorted output power (P_o) to the load (R_L).

Step 10.   Based on the supply voltage (V_{CC}) and the average collector current ($I_{C(AVG)}$), calculate the power supplied by the dc voltage source (P_{DC}).

Step 11.   Based on the power supplied by the dc voltage source (P_{DC}) and the maximum undistorted output power (P_o) calculated in Step 9, calculate the efficiency (η) of the amplifier.

Question: How did the efficiency of this class B push-pull amplifier compare with the efficiency of the class A amplifier in Experiment 20?

__Troubleshooting Problems__

1.   Open circuit file open FIG21-4 and run the simulation. Locate the defective component and state the defect (short or open). You can use any instrument available and make any measurement desired.

        Defective component: _____   Defect: _____

2.   Open circuit file FIG21-5 and run the simulation. Locate the defective component and state the defect (short or open). You can use any instrument available and make any measurement desired.

        Defective component: _____   Defect: _____

3.   Open circuit file FIG21-6 and run the simulation. Locate the defective component and state the defect (short or open). You can use any instrument available and make any measurement desired.

        Defective component: _____   Defect: _____

22

Class C Amplifier

Objectives:

1. Observe the output waveshape of a basic untuned class C common-emitter amplifier with a sine wave input.
2. Determine the duty cycle of the output waveshape of a basic untuned class C common-emitter amplifier with a sine wave input.
3. Draw the dc and ac load lines and locate the operating point (Q-point) for a large-signal class C common-emitter amplifier.
4. Measure the clamped dc voltage on the transistor base of a class C common-emitter amplifier with clamper bias and relate the clamped voltage to the peak value of the input sine wave.
5. Determine the tuned frequency of a class C common-emitter amplifier.
6. Observe the output waveshape of a class C common-emitter amplifier with a sine wave input at the tuned frequency.
7. Measure the undistorted output power for a class C common-emitter amplifier.
8. Determine the bandwidth of a class C common-emitter amplifier.
9. Determine the efficiency of a class C common-emitter amplifier.

Materials:

One 2N3904 bipolar junction transistor
One 10 V dc power supply
Capacitors: two 1 μF, one 0.01 μF
One 1 mH inductor
One function generator
One dual-trace oscilloscope
Resistors: one 10 kΩ, one 20 kΩ, and one 100kΩ

Theory:

When an amplifier is biased below cutoff so that conduction occurs for much less than one-half cycle of the input sine wave (less than 180°), it is classified as a **class C amplifier**. A common method of biasing a class C amplifier below cutoff is with a negative dc base bias **clamping circuit**, as shown in Figure 22-1. The transistor base-emitter junction acts as a diode. When the input goes positive, base capacitor (C_1) charges to the peak value of the input voltage (V_p) minus the base-emitter voltage (0.7 V) through the forward-biased base-emitter diode. If the time constant of the base RC circuit (R_1C_1) is much greater than the time period of the input sine wave, this charged capacitor will stay charged during the input sine wave negative half cycle and produce an average voltage at the base of $V_B = -(V_p - 0.7$ V). This places the transistor below cutoff, except for the positive peaks of the input when the transistor conducts for short time intervals, producing short duration output pulses. Because the transistor collector current is flowing for such a short time duration, less power is dissipated in the transistor, leaving more power for the load. For this reason, a class C amplifier is **more efficient** than a class A or a class B push-pull amplifier. The **duty cycle (D)** of the short duration output pulses can be calculated from

$$D = \frac{W}{T}$$

where W is the transistor *on* time (collector voltage low) and T is the time period for one complete output cycle.

Because the output voltage consists of short duration pulses, a resistance loaded class C amplifier, as shown in Figure 22-1, will not produce a sine wave output when the input is a sine wave. It is therefore necessary to use a class C amplifier with a **parallel resonant tank circuit**, as shown in Figure 22-2. If the parallel resonant circuit (L_1 and C_2) is tuned to the input frequency, the short duration pulses of collector current on each cycle of the input sustains the damped oscillation of the parallel resonant tank circuit so that a sinusoidal output voltage is produced. Because of the parallel resonant circuit, the output voltage is highest when the input is at the resonant frequency and drops off when the input is above or below the resonant frequency. For this reason, class C amplifiers are always intended to amplify a **narrow band** of frequencies and are normally limited to applications as **tuned amplifiers** in radio frequency applications. The **resonant frequency (f_r)** of the parallel resonant LC circuit in Figure 22-2 can be calculated from

$$f_r = \frac{1}{2\pi\sqrt{L_1 C_2}}$$

The Q of the resonant tank circuit is calculated from the Q of the inductor, which is

$$Q = \frac{X_L}{R_S}$$

where $X_L = 2\pi f_r L_1$ and R_S represents the wire resistance of the inductor. The **parallel equivalent resistance** of the resonant tank circuit (R_P) at the resonant frequency can be calculated from

$$R_P = Q X_L$$

and the **impedance (Z)** of the resonant tank circuit at the resonant frequency is equal to the parallel equivalent resistance (R_P).

The **dc load line** for the transistor in Figure 22-2 is approximately a vertical line crossing the horizontal axis at $V_{CE} = V_{CC}$ because the winding resistance (R_S) of the inductor in the resonant circuit is very low. (The slope of the dc load line is the inverse of the dc resistance in the collector circuit). The **Q-point** is on the dc load line at $I_C = 0$ because the transistor is below cutoff. The **ac load line** crosses through the Q-point and crosses the vertical axis at V_{CC}/R_c, where R_c is the ac resistance in the collector circuit. The **ac resistance (R_c)** in the collector circuit is the parallel equivalent of the load resistance (R_L) and the impedance of the resonant tank circuit at the resonant frequency (R_P). Therefore,

$$R_c = \frac{R_P R_L}{R_P + R_L}$$

The quality factor of the ac equivalent output resonant circuit (Q_T) is calculated from

$$Q_T = \frac{R_c}{X_L}$$

The **bandwidth (BW)** of the ac equivalent output resonant circuit (bandwidth of the amplifier) is calculated from

$$BW = \frac{f_r}{Q_T}$$

The **amplifier output power (P_o)** for the circuit in Figure 22-2 can be determined from the peak output voltage (V_o) and the ac resistance (R_c) using the equation

$$P_o = \frac{V_{rms}^2}{R_c} = \frac{(V_o/\sqrt{2})^2}{R_c} = \frac{V_o^2}{2R_c}$$

Because the peak voltage across the parallel resonant circuit is approximately equal to the supply voltage (V_{CC}), the expected **maximum amplifier output power** can be calculated from

$$P_o \cong \frac{V_{CC}^2}{2R_c}$$

The **percent efficiency (η)** of the class C amplifier in Figure 22-2 can be determined from the output power (P_o) and the average power dissipated in the transistor (P_D) using the equation

$$\eta = \frac{P_o}{P_{IN}} = \frac{P_o}{P_o + P_D}$$

The **average power dissipated in the transistor (P_D)** can be determined from the equation

$$P_D = (D)(V_{CE(sat)})(I_{C(sat)})$$

where D is the duty cycle of the short duration output pulses calculated earlier, $V_{CE(sat)}$ is the collector-emitter voltage during the transistor *on* time (collector voltage low), and $I_{C(sat)}$ is the transistor collector current during the transistor *on* time and can be determined from

$$I_{C(sat)} = \frac{V_{CC} - V_{CE(sat)}}{R_c} \cong \frac{V_{CC}}{R_c}$$

Figure 22-1 Untuned Class C Amplifier

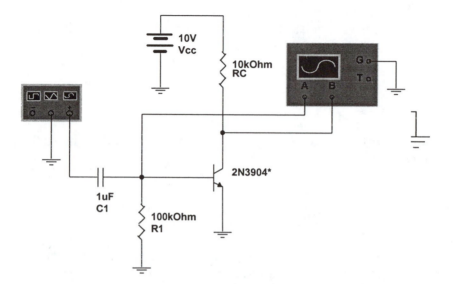

Figure 22-2 Class C Amplifier

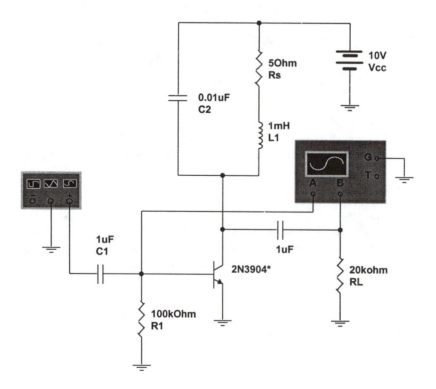

Procedure:

Step 1.   Open circuit file FIG22-1. Bring down the oscilloscope enlargement and make sure that the following settings are selected: Time base (Scale = 5 µs/Div, Xpos = 0, Y/T), Ch A (Scale = 500 mV/Div, Ypos = 0, DC), Ch B (Scale = 5 V/Div, Ypos = 0, DC), Trigger (Pos edge, Level = 0, Nor, A). Bring down the function generator enlargement and make sure that the following settings are selected: *Sine Wave*, Freq = 50 kHz, Ampl = 1 V, Offset = 0. Run the simulation for at least 3 ms transient time, then pause the simulation. You have plotted the amplifier input on the transistor base (red) and output (blue) on the oscilloscope. Draw the output (blue) and transistor base input (red) waveshapes in the space provided. Note the maximum and minimum input and output voltage levels, the output down-time pulse width (W), and the total time period (T) for one complete cycle of the output. **Measure the output voltage minimum and the down-time pulse width accurately using the cursors on the oscilloscope.**

Question: What is the major difference between the output and input waveshapes? **Explain.**

Step 2.   From the output curve plot (blue) in Step 1, calculate the output duty cycle (D).

Step 3.   From the voltages on the transistor base input curve plot (red) in Step 1, determine the clamped dc voltage (average voltage) on the transistor base (V_B).

Step 4.   Based on the peak base input voltage (V_p), calculate the expected clamped dc voltage on the transistor base (V_B).

Questions: Is the transistor base biased above or below transistor cutoff? **Explain.**

What is the relationship between the clamped dc voltage (average voltage) on the transistor base and the peak input signal voltage?

How did the calculated value of the clamped dc voltage on the base in Step 4 compare with the measured value in Step 3?

Step 5.   Open circuit file FIG22-2. Make sure that the instrument settings are the same as in Step 1. Draw the dc load line in the space provided. Based on the circuit in Figure 22-2, locate the operating point (Q-point) on the dc load line.

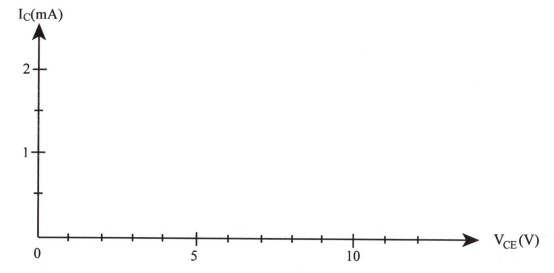

Step 6.   Calculate the resonant frequency (f$_r$) of the parallel resonant circuit in Figure 22-2.

Step 7.   Adjust the frequency of the function generator to the resonant frequency (f$_r$) calculated in Step 6. Run the simulation for at least 2 ms transient time, then pause the simulation. You have plotted the amplifier transistor base input (red) and output (blue) on the oscilloscope. Draw the output (blue) and transistor base input (red) waveshapes in the space provided. *Note the maximum and minimum input and output voltage levels on the curve plot. Also note the time period (T) for one cycle of the output (blue) on the curve plot.*

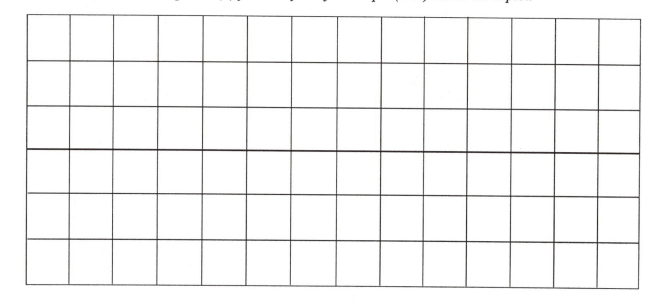

Questions: What is the difference between the output waveshape for the class C amplifier with the parallel resonant circuit compared to the amplifier without the parallel resonant circuit? **Explain why they are different.**

How does the peak output voltage compare with the value of the dc voltage supply (V_{CC})?

Step 8.   Based on the time period (T) for one cycle of the output, calculate the amplifier output frequency (f).

Question: How did the amplifier output frequency compare with the input frequency on the function generator?

Step 9.   Calculate the Q of the parallel resonant tank circuit inductor. (Note: If this experiment is being performed in a hardwired lab, use the actual inductor resistance in place of R_S.)

Step 10.   Calculate the equivalent parallel resistance (R_P) of the resonant tank circuit at the resonant frequency. (See note in Step 9.)

Step 11.   Determine the impedance (Z) of the parallel resonant tank circuit at the resonant frequency.

Step 12.   Determine the equivalent ac resistance (R_c) in the collector circuit at the resonant frequency.

Step 13.   Draw the ac load line on the plot in Step 5.

Question: How does the ac load line compare with the dc load line? **Explain.**

Step 14.   Calculate the quality factor of the ac equivalent output resonant circuit (Q_T).

Question: How does Q_T compare with the Q for the inductor calculated in Step 9? **Explain any difference.**

Step 15.   Calculate the bandwidth (BW) of the ac equivalent output resonant circuit from the value of Q_T and the resonant frequency (f_r).

Step 16.   Determine the ac amplifier output power (P_o) from the ac peak output voltage (V_o) and the ac resistance in the collector circuit (R_c).

Step 17.   Calculate the expected maximum amplifier output power (P_o).

Question: How did the actual output power compare with the expected maximum output power?

Step 18.   Calculate the average power dissipated in the transistor (P_D) from the duty cycle (D).

Step 19.   Calculate the percent efficiency (η) of the amplifier.

Question: How did the efficiency of the class C amplifier compare with the efficiency of the class A and the class B push-pull amplifiers? **Explain.**

Troubleshooting Problems

1. Open circuit file FIG22-3 and run the simulation. Locate the defective component and state the defect (short or open). You can use any instrument available and make any measurement desired.

    Defective component _____   Defect_____

2. Open circuit file FIG22-4 and run the simulation. Locate the defective component and state the defect (short or open). You can use any instrument available and make any measurement desired.

    Defective component _____   Defect_____

3. Open circuit file FIG22-5 and run the simulation. Locate the defective component and state the defect (short or open). You can use any instrument available and make any measurement desired.

    Defective component _____   Defect_____

4. Open circuit file FIG22-6 and run the simulation. Locate the defective component and state the defect (short or open). You can use any instrument available and make any measurement desired.

    Defective component _____   Defect_____

Field-Effect Transistors

The following six experiments involve field-effect transistors. First, you will plot the characteristic curves for the JFET and MOSFET that will be used in the experiments in Part III. Next, you will study different JFET and MOSFET biasing networks and determine their operating point stability. Finally, you will analyze a JFET small-signal common-source amplifier and a MOSFET small-signal common-source amplifier.

The circuits for the experiments in Part III can be found on the enclosed disk in the FET subdirectory.

23

Name_____

Date_____

JFET Characteristics

Objectives:

1. Demonstrate the relationship between drain current, drain-source voltage, and gate-source voltage for a junction field-effect transistor (JFET).
2. Plot the drain characteristic curves for an *n*-channel JFET.
3. Determine the value of the pinch-off voltage for a JFET.
4. Determine the maximum drain current for a JFET.
5. Locate the ohmic region and the constant-current region on the drain characteristic curves.
6. Plot the transfer characteristic curve for an *n*-channel JFET.
7. Determine the gate-source cutoff voltage and compare it to the pinch-off voltage.
8. Determine the JFET forward transconductance from the transfer characteristic curve and show how it depends on the value of the gate-source voltage.
9. Determine the JFET ac drain-source resistance in the constant-current region and the ohmic region of the drain characteristics.

Materials:

One MPF102 *n*-channel JFET
Two dc variable voltage power supplies
Two 0–20 V dc voltmeters
One 0–20 mA dc milliammeter

Theory:

The bipolar junction transistor (BJT), studied in the previous section, depends upon two types of charge carriers called holes and electrons. The **field-effect transistor (FET)** is a **unipolar device** because its operation depends upon only one type of charge carrier. The bipolar junction transistor is a current-controlled device because the collector (output) current is controlled by the base (input) current. The field-effect transistor is a **voltage-controlled device** because the **drain (output) current** is controlled by the **gate (input) voltage**. An FET has a very **high input resistance** compared to a BJT, which has a relatively low input resistance. There are two types of unipolar transistors: the **junction field-effect transistor (JFET)** and the **metal-oxide semiconductor field-effect transistor (MOSFET)**. In this experiment you will study the JFET characteristics, and in the next experiment you will study the MOSFET characteristics.

The ***n*-channel junction field-effect transistor (JFET)** consists of an ***n*-type semiconductor channel** with *p*-type semiconductor diffused into it. Wire leads are connected to each end of the semiconductor channel to form the **drain** (output) and **source**. A wire lead is connected to the p-type semiconductor to form the **gate** (input). The JFET is always operated with the gate-source pn junction reverse-biased. Therefore, the input resistance is very high, giving the JFET a big advantage over the bipolar transistor in situations with very low input signal levels. The reverse-biased pn junction causes a **depletion region** in the channel. The size of the depletion region controls the conduction width of the channel, which controls the drain-source current. The size of the depletion region depends on the gate-source reverse-bias voltage level. Therefore, the drain-source current is controlled by the gate-source voltage.

A p-**channel junction field-effect transistor (JFET)** consists of a p-**type semiconductor channel** with n-type semiconductor diffused into it. The p-channel JFET is identical to the n-channel JFET except that the gate-source voltage is reversed. The schematic symbols for the n-channel and p-channel JFETs indicate the type. For the n-channel JFET symbol the arrow on the gate points toward the channel and for the p-channel JFET symbol the arrow on the gate points away from the channel.

The **JFET drain characteristic curves** can be plotted by recording the drain current (I_D) as a function of drain-source voltage (V_{DS}) for a constant gate-source voltage (V_{GS}), for several V_{GS} values. The **JFET transfer characteristic curve** can be plotted by plotting the drain current (I_D) versus gate-source voltage (V_{GS}) for a constant drain-source voltage (V_{DS}).

The curve plotter circuit shown in Figure 23-1 will be used to plot the drain characteristic curves and the transfer characteristic curve for the MPF102 n-channel JFET. The **pinch-off voltage (V_P)** and the **maximum drain current (I_{DSS})** can be found from the drain characteristic curves. The pinch-off voltage (V_P) is the value of V_{DS} on the $V_{GS} = 0$ curve where the drain current begins to be constant (at the knee) with variations in V_{DS}. The maximum drain current (I_{DSS}) is the value of I_D on the $V_{GS} = 0$ curve where the drain current (I_D) is constant with variations in V_{DS}.

The **ohmic region** of the drain characteristic curves is where there is a large variation in drain current (I_D) for small changes in drain-source voltage (V_{DS}). The **constant-current region** is the part of the curve plot where the drain current is nearly constant for large variations in drain-source voltage (V_{DS}).

The **gate-source cutoff voltage ($V_{GS(off)}$)** is the value of V_{GS} on the transfer characteristic curve that causes the drain current (I_D) to go to zero. The gate-source cutoff voltage ($V_{GS(off)}$) and the pinch-off voltage (V_P) are always equal in magnitude but opposite in sign.

The **JFET forward transconductance (g_m)** equals the change in drain current (ΔI_D) divided by the change in gate-source voltage (ΔV_{GS}) with the drain-source voltage (V_{DS}) constant. The forward transconductance is found on the transfer characteristic curve by drawing the tangent to the curve at a particular value of V_{GS}. The slope of the tangent is the forward transconductance at the value of V_{GS}. Therefore,

$$g_m = \frac{\Delta I_D}{\Delta V_{GS}} = \text{slope}$$

Because the transfer characteristic curve is nonlinear, g_m varies in value as a function of V_{GS}. The value of g_m at $V_{GS} = 0$ is called g_{m0}.

The **ac drain-source (output) resistance (r_{ds})** is found from the drain characteristic curves by measuring the slope at a particular point on one of the curves. The value of the drain-source resistance (r_{ds}) is equal to the inverse of the slope. Therefore,

$$r_{ds} = \frac{\Delta V_{DS}}{\Delta I_D} = \frac{1}{slope}$$

Figure 23-1 JFET Curve Plotter Circuit

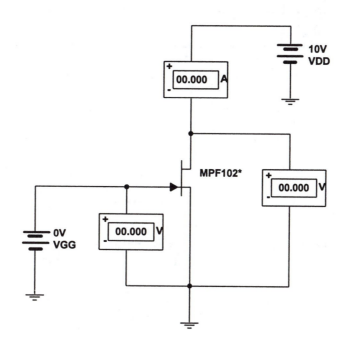

Procedure:

Step 1.   Open circuit file FIG23-1 and run the simulation. Record the reading of the drain current (I_D) in the appropriate location in Table 23-1 based on the drain-source voltage (V_{DS}) and the gate-source voltage (V_{GS}).

Table 23-1 Drain Current (I_D) in mA

V_{GG} (V_{GS}) (Volts)	V_{DD} (V_{DS}) (Volts)						
	0	1	2	3	4	10	20
0							
−1							
−2							
−3							
−4							

Step 2.   Change the value of V_{DD} (V_{DS}) to each value in Table 23-1; run the simulation, then record the drain current (I_D) for each value of V_{DS}.

Step 3.   Change the value of V_{GG} (V_{GS}) to each value in Table 23-1 and follow the procedure in Step 2 until all of the values of the drain current (I_D) are recorded in the table.

Questions: Based on the data in Table 23-1, is a JFET a current-controlled or a voltage-controlled device? **Explain.**

Based on the data in Table 23-1, which is the drain current (I_D) most dependent upon, the drain-source voltage (V_{DS}) or the gate-source voltage (V_{GS})?

Step 4.   Based on the values in Table 23-1, plot the I_D and V_{DS} data points on the graph for each
           value of V_{GS}. Then draw the curve plot for each gate-source voltage (V_{GS}).

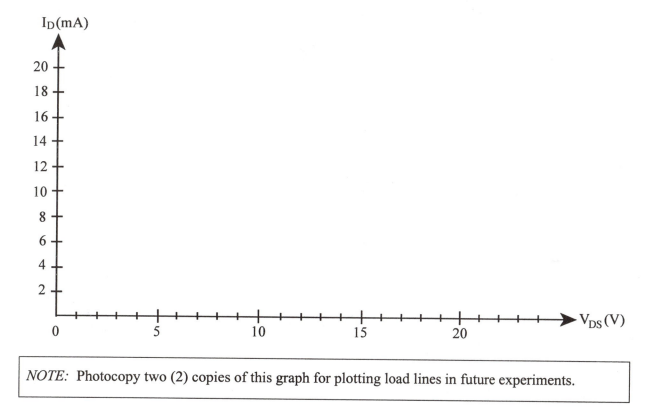

NOTE: Photocopy two (2) copies of this graph for plotting load lines in future experiments.

Step 5.   From the drain characteristics plotted in Step 4, determine the pinch-off voltage (V_P) and
           the maximum drain current (I_{DSS}) at a drain-source voltage (V_{DS}) of 10 V for the JFET.
           Record your answers on the curve plot.

Question: What is the significance of the maximum drain current (I_{DSS}) for a junction field-effect
transistor?

Step 6.   Note the ohmic region and the constant-current region on the drain characteristics plotted
           in Step 4.

Step 7.   Based on the values of the drain current (I_D) and gate-source voltage (V_{GS}) recorded in
              Table 23-1 for $V_{DS} = 10$ V, plot the transfer characteristic curve (I_D vs V_{GS}).

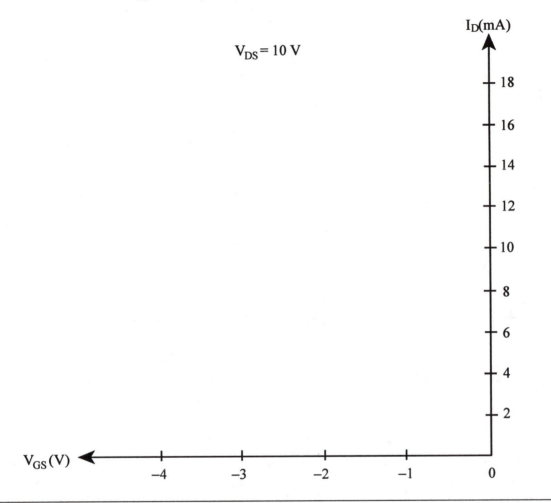

$V_{DS} = 10$ V

I_D(mA)

18

16

14

12

10

8

6

4

2

V_{GS}(V)

−4   −3   −2   −1   0

NOTE: Photocopy two (2) copies of this graph for plotting load lines in future experiments.

Step 8.   Based on the transfer characteristic curve plotted in Step 7, determine the value of the
              gate-source cutoff voltage ($V_{GS(off)}$) and record your answer on the curve plot.

Questions: How does the pinch-off voltage (V_P) compare with the gate-source cutoff voltage ($V_{GS(off)}$)?

What is the significance of $V_{GS(off)}$ and V_P for a junction field-effect transistor?   What is the range of
V_{GS} for this JFET? **Explain.**

Step 9.   Based on the transfer characteristic curve plotted in Step 7, determine the value of the JFET forward transconductance (g_m) at $V_{GS} = -1$ V. Record your answer. *Hint:* Measure the slope of the tangent to the curve at $V_{GS} = -1$ V.

$g_m =$ _____ at $V_{GS} = -1$ V

Step 10.   Following the same procedure, find g_m at $V_{GS} = -2$ V.

$g_m =$ _____ at $V_{GS} = -2$ V

Question: Is the JFET transconductance (g_m) constant or a function of V_{GS}? **Explain.**

Step 11.   From the drain characteristics plotted in Step 4, determine the ac drain-source resistance (r_{ds}) in the constant-current region. Record your answer. *Hint:* Measure the slope of the curve at $V_{GS} = 0$, $V_{DS} = 10$ V.

$r_{ds} =$ _____ in the constant-current region

Step 12.   From the drain characteristics plotted in Step 4, determine the ac drain-source resistance (r_{ds}) in the ohmic region. Record your answer. *Hint:* Measure the slope of the curve at $V_{GS} = 0$ V, $V_{DS} = 2$ V.

$r_{ds} =$ _____ in the ohmic region

Questions: Is the JFET ac drain-source resistance (r_{ds}) large or small in the constant-current region of the drain characteristics? In the ohmic region? How did the value of r_{ds} compare in the two regions? **Explain.**

EXPERIMENT

24

MOSFET Characteristics

Name_____

Date_____

Objectives:

1. Demonstrate the relationship between drain current, drain-source voltage, and gate-source voltage for a depletion MOSFET (D-MOSFET) and an enhancement MOSFET (E-MOSFET).
2. Plot the drain characteristic curves for an *n*-channel D-MOSFET and an *n*-channel E-MOSFET.
3. Determine the value of the pinch-off voltage for a D-MOSFET.
4. Determine the zero-bias drain current for a D-MOSFET.
5. Plot the transfer characteristic curves for a D-MOSFET and an E-MOSFET.
6. Determine the D-MOSFET gate-source cutoff voltage and compare it to the pinch-off voltage.
7. Determine the gate-source threshold voltage for an E-MOSFET.
8. Determine the D-MOSFET forward transconductance from the transfer characteristic curve.
9. Determine the difference between the depletion mode and the enhancement mode for a D-MOSFET.
10. Determine the difference between the D-MOSFET and E-MOSFET characteristic curves.

Materials:

One 2N3796 *n*-channel D-MOSFET
One IRF511 *n*-channel E-MOSFET
Two variable voltage dc power supplies
Two 0–20 V dc voltmeters
One 0–10 mA dc milliammeter
One 0–500 mA dc milliameter

Theory:

The **metal oxide semiconductor field-effect transistor (MOSFET)** has a **drain**, **gate**, and **source** but does not have a *pn* junction between the gate and source like the junction field-effect transistor (JFET). The MOSFET gate is **insulated** from the channel by a **silicon dioxide (SiO$_2$) layer**. This gives the MOSFET a higher input resistance than the JFET. There are two types of MOSFETs, the **depletion type (D-MOSFET)** and the **enhancement type (E-MOSFET)**.

The **D-MOSFET** consists of a narrow *n*-type or *p*-type semiconductor channel with silicon dioxide deposited adjacent to it and a metal surface to form the gate. The silicon dioxide insulates the gate from the channel. Wire leads are connected to the **gate** (input) and to each end of the semiconductor channel to form the **drain** (output) and **source**. The *p*-channel and *n*-channel operation is the same, except the voltage polarities are reversed. We will use an *n*-channel D-MOSFET in this experiment.

Because the gate is insulated from the channel, the gate can have either a positive or negative voltage polarity. Therefore, the D-MOSFET can operate in one of two modes. When the gate-source voltage is negative, the n-channel D-MOSFET is in the **depletion mode**, and when the gate-source voltage is positive, it is in the **enhancement mode**. For this reason it is sometimes called an enhancement/depletion MOSFET. When a negative voltage is applied to the gate, the number of negative charge carriers in the channel is reduced, lowering the channel conductivity. The greater the negative voltage on the gate, the lower the channel conductivity and the lower the drain current. At a sufficiently negative gate-source voltage, the channel will be depleted of charge carriers and the drain current will be zero. This negative gate-source voltage is called the **gate-source cutoff voltage ($V_{GS(off)}$)**. Like the n-channel JFET, the n-channel D-MOSFET conducts drain current for negative gate-source voltages between $V_{GS(off)}$ and zero in the depletion mode. When a positive voltage is applied to the gate in the enhancement mode, the n-channel is enhanced with negative charge carriers, increasing the channel conductivity and increasing the drain current. The greater the positive voltage on the gate, the higher the channel conductivity and the higher the drain current.

The **E-MOSFET** is similar to the D-MOSFET, except it has **no channel**. Both the drain and source are n-type semiconductors with p-type semiconductor between them. The p-type semiconductor has silicon dioxide deposited adjacent to it and a metal surface to form the gate. The silicon dioxide insulates the gate from the p-type semiconductor. The pn junctions prevent drain-source current until a positive voltage is applied to the gate. A positive gate voltage above a **threshold voltage ($V_{GS(th)}$)** induces an n-channel in the p region by creating a layer of negative charge carriers. Increasing the positive gate-source voltage increases the conductivity of the channel, increasing the drain current. When the gate-source voltage is below the threshold voltage ($V_{GS(th)}$), there is no drain current. Because the E-MOSFET only conducts drain current when the gate is positive, it operates only in the **enhancement mode** and it has **no depletion mode**. The E-MOSFET is most often used in switching applications because it switches easily between the *on* and *off* states. For this reason it is commonly seen in **digital applications**.

The D-MOSFET and E-MOSFET **drain characteristic curves** can be plotted by recording the drain current (I_D) as a function of drain-source voltage (V_{DS}) for a constant gate-source voltage (V_{GS}), and then plotting the I_D versus V_{DS} curve plots for several V_{GS} values. The D-MOSFET and E-MOSFET **transfer characteristic curves** can be plotted by plotting the drain current (I_D) versus gate-source voltage (V_{GS}) for a constant drain-source voltage (V_{DS}).

The D-MOSFET and E-MOSFET curve plotter circuits shown in Figures 24-1 and 24-2 will be used to plot the drain characteristic curves and the transfer characteristic curves for the 2N3796 D-MOSFET and the IRF511 E-MOSFET. The **pinch-off voltage (V_P)** and the **zero bias drain current (I_{DSS})** for a D-MOSFET can be determined from the drain characteristic curve plot. The pinch-off voltage (V_P) is the value of V_{DS} on the $V_{GS} = 0$ curve where the drain current begins to be constant (at the knee). The zero bias drain current (I_{DSS}) is the value of I_D on the $V_{GS} = 0$ curve where the drain current (I_D) is constant.

The **gate-source cutoff voltage ($V_{GS(off)}$)** for the D-MOSFET is the value of V_{GS} on the transfer characteristic curve that causes the drain current (I_D) to go to zero. The gate-source cutoff voltage ($V_{GS(off)}$) and the **pinch-off voltage (V_P)** are always equal in magnitude but opposite in sign.

The **gate-source threshold voltage ($V_{GS(th)}$)** for the E-MOSFET is the value of V_{GS} on the transfer characteristic curve where the drain current (I_D) goes to zero.

The D-MOSFET **forward transconductance (g_m)** equals the change in drain current (ΔI_D) divided by the change in gate-source voltage (ΔV_{GS}) with the drain-source voltage (V_{DS}) constant. The forward transconductance is determined from the transfer characteristic curve by drawing the tangent to the curve at a particular value of V_{GS}. The slope of the tangent is the forward transconductance at the value of V_{GS}. Therefore,

$$g_m = \frac{\Delta I_D}{\Delta V_{GS}} = \text{slope}$$

Figure 24-1 D-MOSFET Curve Plotter Circuit

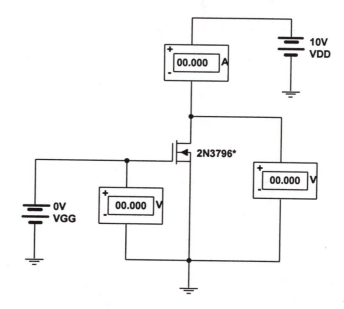

Figure 24-2 E-MOSFET Curve Plotter Circuit

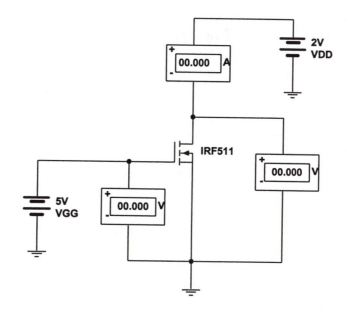

Procedure:

Step 1.   Open circuit file FIG24-1 and run the simulation. Record the reading of the drain current (I_D) in the appropriate location in Table 24-1 based on the drain-source voltage (V_{DS}) and the gate-source voltage (V_{GS}).

Table 24-1 Drain Current (I_D) in mA

V_{GG} (V_{GS}) (Volts)	\multicolumn{7}{c}{V_{DD} (V_{DS}) (Volts)}
	0	1	2	3	4	10	20
+2							
+1							
0							
−1							
−2							
−3							

Step 2.   Change the value of V_{DD} (V_{DS}) to each value in Table 24-1, then run the simulation and record each drain current (I_D) in the table.

Step 3.   Change the value of V_{GG} (V_{GS}) to the next value in Table 24-1 and follow the procedure in Step 2 until all of the values of the drain current (I_D) are recorded in the table for each value of V_{GS}.

Question: Based on the data collected, which is the drain current (I_D) most dependent upon, the drain-source voltage (V_{DS}) or the gate-source voltage (V_{GS})?

Step 4.   Based on the values in Table 24-1, plot the I_D and V_{DS} data points on the graph for each
               value of V_{GS}. Then draw the curve plot for each gate-source voltage (V_{GS}).

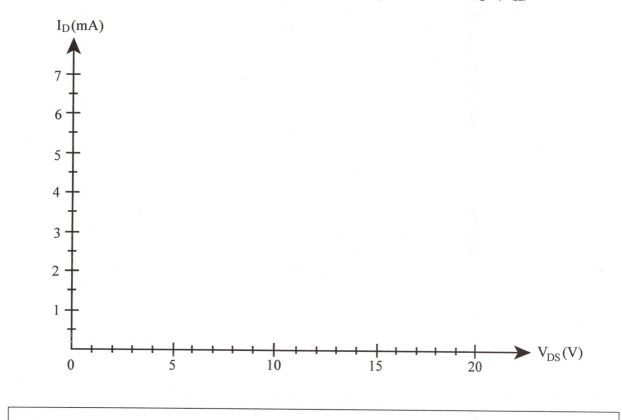

NOTE: Make two (2) photocopies of this graph for plotting load lines in future experiments.

Step 5.   From the drain characteristics plotted in Step 4, determine the pinch-off voltage (V_P) and
               the zero bias drain current (I_{DSS}) for the D-MOSFET. Record your answers on the graph.

Question: What is the significance of the zero bias drain current (I_{DSS}) for a D-MOSFET?

Step 6.   Based on the values of the drain current (I_D) and gate-source voltage (V_{GS}) recorded in Table 24-1 for $V_{DS} = 10$ V, plot the transfer characteristic curve (I_D vs V_{GS}).

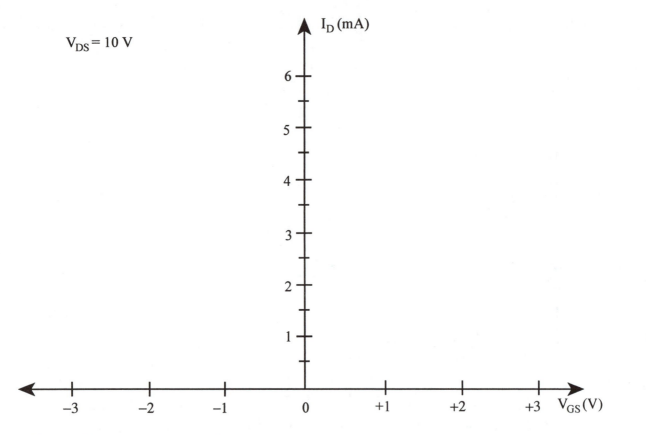

Step 7.   Based on the transfer characteristic curve plotted in Step 6, determine the value of the D-MOSFET gate-source cutoff voltage ($V_{GS(off)}$) and record your answer on the graph.

Questions: How does the pinch-off voltage (V_P) compare with the gate-source cutoff voltage ($V_{GS(off)}$) for the D-MOSFET?

What is the significance of $V_{GS(off)}$ for a D-MOSFET?

Step 8.   Based on the transfer characteristic curve plotted in Step 6, determine the value of the D-MOSFET forward transconductance (g_m) at $V_{GS} = 0$ V and record your answer.
Hint: Measure the slope of the tangent to the curve at $V_{GS} = 0$ V.

g_m = _____ at $V_{GS} = 0$ V

Questions: What is the major difference between the characteristics for the D-MOSFET compared to the JFET?

What is the major difference between the depletion mode and the enhancement mode for the D-MOSFET?

Step 9.   Open circuit file FIG24-2 and run the simulation. Record the reading of the drain current (I_D) in the appropriate location in Table 24-2 based on the drain-source voltage (V_{DS}) and the gate-source voltage (V_{GS}).

Table 24-2 Drain Current (I_D) in mA

V_{GG} (V_{GS}) (Volts)	V_{DD} (V_{DS}) (Volts)						
	0	1	2	3	4	10	20
+7							
+6							
+5							
+4							
+3							

Step 10.   Change the value of V_{DD} (V_{DS}) to each value in Table 24-2, then run the simulation and record each drain current (I_D) in the table.

Step 11.   Change the value of V_{GG} (V_{GS}) to the next value in Table 24-2 and follow the procedure in Step 10 until all of the values for drain current (I_D) are recorded in the table for each value of V_{GS}.

Question: Based on the data collected, which is the drain current (I_D) most dependent upon, the drain-source voltage (V_{DS}) or the gate-source voltage (V_{GS})?

Step 12.   Based on the values in Table 24-2, plot the I_D and V_{DS} data points on the graph for each value of V_{GS}. Then draw the curve plot for each gate-source voltage (V_{GS}).

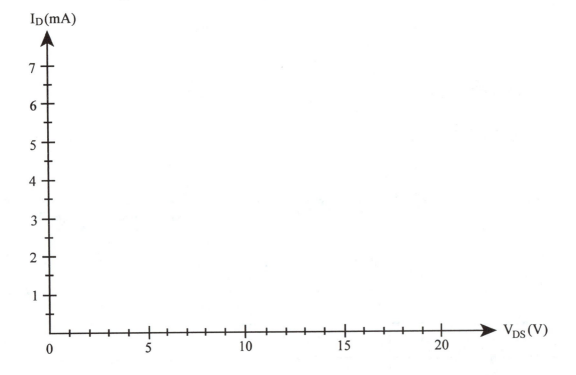

NOTE: Photocopy two (2) copies of this graph for plotting load lines in future experiments.

Step 13.   Based on the values of the drain current (I_D) and gate-source voltage (V_{GS}) recorded in Table 24-2 for $V_{DS} = 10$ V, plot the transfer characteristic curve (I_D vs V_{GS}).

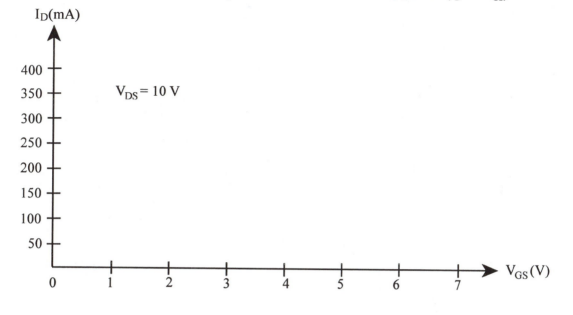

NOTE: Make two (2) photocopies of this graph for plotting load lines in future experiments.

Step 14.   Based on the transfer characteristic curve plotted in Step 13, determine the gate-source threshold voltage ($V_{GS(th)}$). Record your answer on the graph.

Questions: Is a MOSFET a current-controlled or a voltage-controlled device? **Explain.**

What is the significance of the gate-source threshold voltage ($V_{GS(th)}$) for an E-MOSFET?

What is the major difference between the characteristic curves for the D-MOSFET and the E-MOSFET?

What is the most common application for an E-MOSFET? **Explain why.**

25

JFET Bias Circuits

Name_____

Date_____

Objectives:

1. Study the JFET self-bias circuit.
2. Study the JFET voltage-divider bias circuit.
3. Draw the dc load line on the JFET characteristics for a JFET self-bias circuit, and then locate the operating point (Q-point).
4. Estimate the values of the gate-source voltage, drain current, and drain-source voltage for a JFET self-bias circuit; compare the estimated values with the measured values.
5. Determine the bias stability for a JFET self-bias circuit.
6. Draw the dc load line on the JFET characteristics for a JFET voltage-divider bias circuit, and then locate the operating point (Q-point).
7. Estimate the values of the gate-source voltage, drain current, and drain-source voltage for a JFET voltage-divider bias circuit; compare the estimated values with the measured values.
8. Determine the bias stability for a JFET voltage-divider bias circuit.

Materials:

One MPF102 JFET
One 0–20 V dc power supply
Three 0–15 V dc voltmeters
One 0–10 mA dc milliammeter
Resistors (one each): 100 Ω, 1 kΩ, 1.3 kΩ, 500 kΩ, 1 MΩ, 3.5 MΩ

Theory:

The purpose of **biasing** a JFET is to establish a dc drain current that places the **operating point (Q-point)** as close as possible to the **middle** of the **transfer characteristic curve** where $I_D = I_{DSS}/2$. This allows the maximum amount of drain current variation with **minimum distortion** of the output waveshape. The JFET gate-source junction must always be reverse-biased. (See Experiment 23). For an **n-channel JFET**, this means that the **gate-source voltage (V_{GS})** must always be **negative**. You will study the two major types of JFET bias circuits for accomplishing these objectives, self-bias and voltage-divider bias.

An n-channel JFET **self-bias** circuit is shown in Figure 25-1. Because of the reverse-biased gate-source junction, the gate current is practically zero. Therefore, the voltage across the gate resistor (R_G) is practically zero, making the gate voltage (V_G) practically zero. This makes the dc gate-source voltage (V_{GS}) equal to the negative of the source voltage (V_S), as shown.

$$V_{GS} = V_G - V_S = 0 - V_S = -V_S$$

where $V_S = I_D R_S$. Therefore,

$$V_{GS} = -I_D R_S$$

The **dc load line** is located on the **transfer characteristic curve** plot by first calculating the value of V_{GS} when the drain current is equal to I_{DSS}. This value of V_{GS} can be found from

$$V_{GS} = -I_D R_S = -I_{DSS} R_S$$

Draw the load line on the transfer characteristic curve plot between the origin ($V_{GS} = 0$, $I_D = 0$) and the point where V_{GS} is equal to the value calculated above and I_D is equal to I_{DSS}. The point where this load line crosses the transfer characteristic curve is the **operating point (Q-point)**.

An *n*-channel JFET **voltage-divider bias** circuit is shown in Figure 25-2. The dc **gate-source voltage (V_{GS})** is calculated by subtracting the source voltage from the gate voltage. Assuming that the gate current is zero, the gate-source voltage (V_{GS}) is calculated as follows:

$$V_{GS} = V_G - V_S$$

where $V_S = I_D R_S$ and $V_G = V_{DD}(R_2)/(R_1 + R_2)$

The **dc load line** is located on the **transfer characteristic curve** plot by first calculating the value of V_{GS} when the drain current (I_D) is equal to zero. This value of V_{GS} can be determined from

$$V_S = I_D R_S = (0)R_S = 0$$

$$V_{GS} = V_G - V_S = V_G - 0 = V_G$$

This is the load line crossing point on the horizontal axis. Next calculate the value of I_D when the gate-source voltage (V_{GS}) is zero. This value of I_D can be determined from

$$I_D = \frac{V_S}{R_S} = \frac{V_G - V_{GS}}{R_S} = \frac{V_G - 0}{R_S} = \frac{V_G}{R_S}$$

This is the load line crossing point on the vertical axis. The point where this load line crosses the transfer characteristic curve is the **operating point (Q-point)**.

Because the voltage-divider bias load line on the transfer characteristic curve plot has a lower slope than the self-bias load line, there is less variation in drain current (I_D) with variations in JFET transfer characteristics. Therefore, **voltage-divider bias** provides **better operating point stability** than self-bias.

For the self-biased JFET amplifier circuit in Figure 25-1 and the voltage-divider biased JFET amplifier circuit in Figure 25-2, the **dc load line** crosses the horizontal axis on the **drain characteristic curve plot** at a value of drain-source voltage (V_{DS}) equal to V_{DD}. The dc load line crosses the vertical axis at a value of drain current (I_D) equal to $V_{DD}/(R_D + R_S)$. This drain current is approximately equal to the **JFET saturation current ($I_{D(sat)}$)**.

Figure 25-1 JFET Self-Bias Circuit

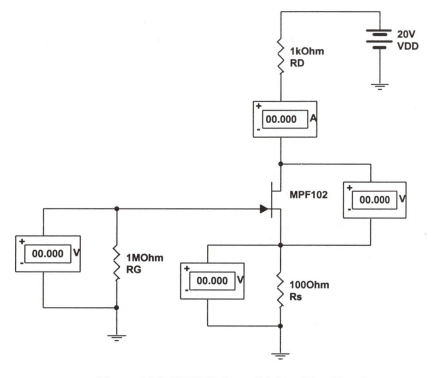

Figure 25-2 JFET Voltage-Divider Bias Circuit

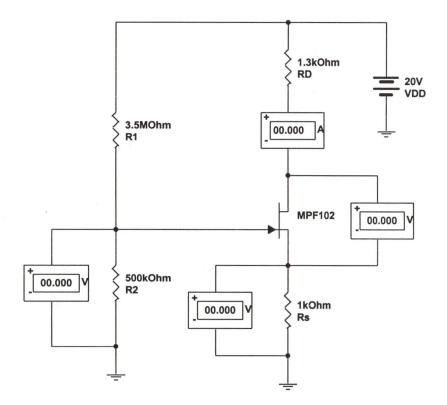

Procedure:

Step 1.   Open circuit file FIG25-1 and run the simulation. Record the drain current (I_D), the drain-source voltage (V_{DS}), the source voltage (V_S), and the gate voltage (V_G).

I_D = _____ V_{DS} = _____ V_S = _____

V_G = _____

Step 2.   Based on the readings in Step 1, calculate the value of the gate-source voltage (V_{GS}).

Step 3.   Based on the circuit values in Figure 25-1, draw the dc load line on a copy of the MPF102 drain characteristics and transfer characteristics plotted in Experiment 23, Steps 4 and 7.

Step 4.   Based on where the dc load line crosses the transfer characteristic curve, locate the Q-point. From the Q-point estimate the value of the gate-source voltage (V_{GS}) and record your answer. Based on that value, locate the Q-point on the drain characteristic curve plot.

V_{GS} = _____

Questions: How did your estimated gate-source voltage (V_{GS}) compare with the measured value in Steps 1-2?

Was the Q-point for the self-biased JFET near the middle of the load line on the drain characteristics?

Step 5.   Based on the Q-point located on the drain characteristic curves, estimate the values of the drain current (I_D) and the drain-source voltage (V_{DS}). Record your answers.

I_D = _____ V_{DS} = _____

Question: How did your estimated drain current (I_D) and drain-source voltage (V_{DS}) compare with the measured values in Step 1?

Step 6.   Double click the JFET and click *Edit Model*. Change the transconductance coefficient (Beta) from 1 m to 0.5 m; then click *Change Part Model*. Click *OK* on the Models window to return to the circuit. You have changed the transconductance coefficient for the MPF102. This will allow you to examine the effect that changing JFETS would have on the dc drain current (Q-point bias stability) for a self-biased JFET circuit. Run the simulation and record the new values of drain current (I_D) and drain-source voltage (V_{DS}).

I_D = _____ V_{DS} = _____

> *NOTE:* If you are using this manual in a lab environment, you should change the MPF102 JFET instead of changing the JFET parameters.

Step 7.   Based on the new values for V_{DS} and I_D, locate the new Q-point on the load lines drawn in Step 3. (The characteristic curves no longer have the same gate-source voltages because changing the JFET characteristics moved the curves).

Step 8.   Double click the JFET, click *Edit Model*, and return the value of the transconductance coefficient (Beta) back to 1 m for the MPF102 JFET. Don't forget to click *Change Part Model*, and *OK* in both windows.

> *NOTE:* If you are using this manual in a lab environment, you should change back to the original MPF102 JFET.

Step 9.   Open circuit file FIG25-2 and run the simulation. Record the drain current (I_D), the drain-source voltage (V_{DS}), the source voltage (V_S), and the gate voltage (V_G).

I_D = _____ V_{DS} = _____ V_S = _____

V_G = _____

Step 10   Based on the readings in Step 9, calculate the gate-source voltage (V_{GS}).

Step 11.   Based on the circuit values in Figure 25-2, draw the dc load line on a copy of the MPF102 drain characteristics plotted in Experiment 23, Step 4.

Step 12.   Based on the circuit values in Figure 25-2, calculate the value of I_D when $V_{GS} = 0$ and the value of V_{GS} when $I_D = 0$. Based on these two points, draw the dc load line on a copy of the transfer characteristics plotted in Experiment 23, Step 7.

Step 13.   Based on where the dc load line crosses the transfer characteristic curve, locate the Q-point. From the Q-point estimate the value of the gate-source voltage (V_{GS}) and record your answer. Based on that value, locate the Q-point on the drain characteristic curve plot. From this Q-point, estimate the values of the drain current (I_D) and the drain-source voltage (V_{DS}) and record the values.

V_{GS} = _____ I_D = _____ V_{DS} = _____

Questions: How did your estimated gate-source voltage (V_{GS}), drain current (I_D), and drain-source voltage (V_{DS}) compare with the measured values in Steps 9-10?

Was the Q-point for the voltage-divider biased JFET near the middle of the load line on the drain characteristics?

Step 14.   Double click the JFET and change the value of the transconductance coefficient (Beta) from 1 m to 0.5 m following the procedure in Step 6. Run the simulation and record the value of drain current (I_D) and drain-source voltage (V_{DS}).

I_D = _____ V_{DS} = _____

NOTE: If you are using this manual in a lab environment, you should change the MPF102 JFET instead of changing the JFET parameters.

Step 15.   Based on the new values for V_{DS} and I_D, locate the new Q-point on the load lines drawn in Steps 11-12. (The characteristic curves no longer have the correct gate-source voltages because changing the JFET characteristics moved the curves).

Experiment 25
199

Question: How did drain current stability for the voltage-divider biased JFET compare with drain current stability for the self-biased JFET? Which was more stable? **Explain.**

Step 16.   Double click the JFET, click *Edit Model*, and return the value of the transconductance coefficient (Beta) back to 1 m for the MPF102 JFET. Don't forget to click *Change Part Model*, and *OK* in both windows.

NOTE: If you are using this manual in a lab environment, you should change back to the original MPF102 JFET.

Troubleshooting Problems

1.   Open circuit file FIG25-3 and run the simulation. Based on the measured voltages and currents, what is wrong with the MPF102 JFET?

2.   Open circuit file FIG25-4 and run the simulation. Determine which circuit component is defective and state the defect (open or short) based on the voltage and current readings.

   Defective component: _____   Defect: _____

3.   Open circuit file FIG25-5 and run the simulation. Based on the measured voltages and currents, what is wrong with the MPF102 JFET?

4.   Open circuit file FIG25-6 and run the simulation. Determine which circuit component is defective and state the defect (open or short) based on the voltage and current readings.

      Defective component: _____   Defect: _____

5.   Open circuit file FIG25-7 and run the simulation. Determine which circuit component is defective and state the defect (open or short) based on the voltage and current readings.

      Defective component: _____   Defect: _____

26

Name_____

Date_____

MOSFET Bias Circuits

Objectives:

1. Study a D-MOSFET zero-bias circuit.
2. Study an E-MOSFET voltage-divider bias circuit.
3. Study an E-MOSFET drain-feedback bias circuit.
4. Draw the dc load line on the D-MOSFET characteristics for a D-MOSFET zero-bias circuit and locate the Q-point on the load line.
5. Estimate the values of the drain current and the drain-source voltage for a D-MOSFET zero-bias circuit and compare the estimated values with the measured values.
6. Draw the dc load line on the E-MOSFET characteristics for an E-MOSFET voltage-divider bias circuit and locate the Q-point on the load line.
7. Estimate the values of the drain current, the drain-source voltage, and the gate-source voltage for an E-MOSFET voltage-divider bias circuit and compare the estimated values with the measured values.
8. Determine the bias stability for an E-MOSFET voltage-divider bias circuit.
9. Draw the dc load line on the E-MOSFET characteristics for an E-MOSFET drain-feedback bias circuit and locate the Q-point on the load line.
10. Determine the bias stability for an E-MOSFET drain-feedback bias circuit.

Materials:

One 2N3796 n-channel D-MOSFET
One IRF511 n-channel E-MOSFET
One 0–20 V dc power supply
Two 0–20 V dc voltmeters
One 0–5 mA dc milliammeter
One 0–200 mA dc milliammeter
Resistors (one each): 30 Ω, 50 Ω, 4 kΩ, 100 kΩ, 230 kΩ, 1 MΩ

Theory:

The purpose of **biasing** a MOSFET is to establish a dc drain current that places the **operating point (Q-point)** as close as possible to the middle of the load line. This allows the maximum amount of drain current variation with **minimum distortion** of the output waveshape when an input signal is applied. The **D-MOSFET gate-source voltage** can be **positive or negative**, depending on whether it is operating in the **depletion** or **enhancement** mode. A simple D-MOSFET biasing method is to set **the gate-source** (V_{GS}) to **zero** so that the ac signal varies the gate-source voltage above and below the bias voltage. An

n-channel **E-MOSFET gate-source voltage** must always be **positive** and above the **gate-source threshold voltage ($V_{GS(th)}$)**. Therefore, an E-MOSFET gate-source voltage cannot be biased at zero. You will study the three major types of MOSFET bias circuits for accomplishing these objectives, D-MOSFET zero-bias, E-MOSFET voltage-divider bias, and E-MOSFET drain-feedback bias.

An n-**channel D-MOSFET zero-bias circuit** is shown in Figure 26-1. The **gate-source voltage (V_{GS})** is calculated based on the assumption that the gate current is zero, making the voltage across the gate resistor (R_G) equal zero. This makes the gate voltage (V_G) equal zero. Because the source is grounded, the gate-source voltage (V_{GS}) is determined from

$$V_{GS} = V_G - V_S = V_G - 0 = V_G = 0$$

For the D-MOSFET zero-biased circuit in Figure 26-1, the **dc load line** crosses the **horizontal axis** on the drain characteristic curve plot at a value of drain-source voltage (V_{DS}) equal to V_{DD} and crosses the **vertical axis** at a value of drain current (I_D) equal to V_{DD} / R_D. This value of I_D is approximately equal to the **saturation current $I_{D(sat)}$**. The **operating point (Q-point)** is located on the dc load line where the load line crosses the $V_{GS} = 0$ curve.

An n-**channel E-MOSFET voltage-divider bias circuit** is shown in Figure 26-2. The **gate-source voltage** is calculated by subtracting the source voltage (V_S) from the gate voltage (V_G). Because the gate current is assumed to be zero and the source is grounded, the gate-source voltage (V_{GS}) is determined from

$$V_{GS} = V_G - V_S = V_G - 0 = V_G$$

where $V_G = V_{DD} R_2 / (R_1 + R_2)$ from the voltage-divider rule.

For the E-MOSFET voltage-divider bias circuit in Figure 26-2, the **dc load line** crosses the **horizontal axis** on the drain characteristic curve plot at a value of drain-source voltage (V_{DS}) equal to V_{DD} and crosses the **vertical axis** at a value of drain current (I_D) equal to V_{DD} / R_D. The **operating point (Q-point)** is located on the dc load line where the load line crosses the $V_{GS} = V_G$ curve.

An n-**channel E-MOSFET drain-feedback bias** circuit is shown in Figure 26-3. The **gate-source voltage** is calculated by subtracting the source voltage (V_S) from the gate voltage (V_G). Because the gate current is assumed to be zero, the voltage across the gate resistor (R_G) is zero. This makes the gate voltage (V_G) equal to the drain voltage (V_D). Because the source is grounded, the gate-source voltage (V_{GS}) is equal to the drain-source voltage (V_{DS}) as follows.

$$V_{GS} = V_G - V_S = V_G - 0 = V_G = V_D = V_{DS}$$

For the E-MOSFET drain-feedback biased circuit in Figure 26-3, the **dc load line** crosses the **horizontal axis** on the drain characteristic curve plot at a value of drain-source voltage (V_{DS}) equal to V_{DD} and crosses the **vertical axis** at a value of drain current (I_D) equal to V_{DD} / R_D. The **operating point (Q-point)** is located on the dc load line where the load line crosses the $V_{GS} = V_{DS}$ curve.

Drain-feedback bias is **more stable** than voltage-divider bias for the following reasons. When the MOSFET parameters change due to temperature or device changes, the drain current (I_D) will try to change. This will cause the drain-source voltage (V_{DS}) to change. If the drain current tries to increase, the drain-source voltage (V_{DS}) will drop. With drain-feedback bias, this will reduce the gate-source voltage (V_{GS}), causing the drain current to decrease back toward the original value.

Figure 26-1 D-MOSFET Zero-Bias Circuit

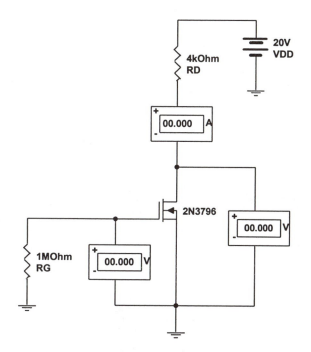

Figure 26-2 E-MOSFET Voltage-Divider Bias Circuit

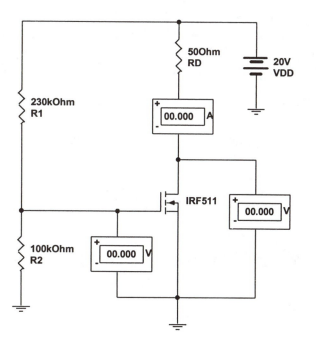

Figure 26-3 E-MOSFET Drain-Feedback Bias Circuit

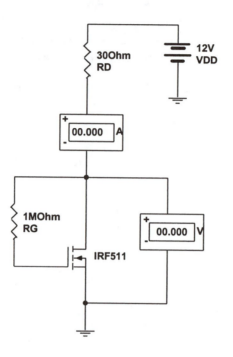

Procedure:

Step 1.   Open circuit file FIG26-1 and run the simulation. Record the drain current (I_D), the drain-source voltage (V_{DS}), and the gate-source voltage (V_{GS}).

I_D = _____ V_{DS} = _____ V_{GS} = _____

Step 2.   Based on the circuit values in Figure 26-1, draw the dc load line on a copy of the 2N3796 D-MOSFET drain characteristics plotted in Experiment 24, Step 4. Based on the value of V_{GS} determined in Step 1, locate the Q-point on the load line. Based on the Q-point, estimate the value of the drain current (I_D) and the drain-source voltage (V_{DS}) and record your answers.

I_D = _____ V_{DS} = _____

Questions: How did your estimated drain current (I_D) and drain-source voltage (V_{DS}) compare with the measured values in Step 1 for the zero-biased D-MOSFET?

Was the Q-point for the zero-biased D-MOSFET near the middle of the load line?

Step 3.   Open circuit file FIG26-2 and run the simulation. Record the drain current (I_D), the drain-source voltage (V_{DS}), and the gate-source voltage (V_{GS}).

I_D = _____ V_{DS} = _____ V_{GS} = _____

Step 4.   Based on the circuit values in Figure 26-2, draw the dc load line on a copy of the IRF511 E-MOSFET drain characteristics plotted in Experiment 24, Step 12.

Step 5.   Calculate the value of V_{GS} from the circuit component values. Based on this value of V_{GS}, locate the Q-point on the load line plotted in Step 4.

Questions: How did your calculated gate-source voltage (V_{GS}) compare with the measured value in Step 3 for the voltage-divider biased E-MOSFET?

Was the Q-point for the voltage-divider biased E-MOSFET near the middle of the load line?

Step 6.   Based on the location of the Q-point, estimate the value of the drain current (I_D) and the drain-source voltage (V_{DS}) and record your answers.

I_D = _____ V_{DS} = _____

Question: How did your estimated drain current (I_D) and drain-source voltage (V_{DS}) compare with the measured values in Step 3 for the voltage-divider biased E-MOSFET?

Step 7.   Following the procedure in Experiment 25, Step 6, change the IRF511 MOSFET transconductance coefficient (KP) from 0.042 to 0.02. Run the simulation and record the values of drain current (I_D) and drain-source voltage (V_{DS}).

I_D = _____ V_{DS} = _____

NOTE: If you are using this manual in a lab environment, you should change the IRF511 MOSFET instead of changing the MOSFET parameters.

Step 8.   Based on the values measured in Step 7, locate the new Q-point on the load line plotted in Step 4.

Step 9.   Return the IRF511 MOSFET transconductance coefficient (KP) back to 0.042.

NOTE: If you are using this manual in a lab environment, you should change back to the original
IRF511 MOSFET.

Step 10.   Open circuit file FIG26-3 and run the simulation. Record the drain current (I_D) and the drain-
               source voltage (V_{DS}).

               I_D = _____ V_{DS} = _____

Step 11.   Based on the values measured in Step 10, calculate the gate-source voltage (V_{GS}).

Step 12.   Based on the circuit values in Figure 26-3, draw the dc load line on a copy of the IRF511
               E-MOSFET drain characteristics plotted in Experiment 24, Step 12.

Step 13.   Based on the value of V_{GS} calculated in Step 11, locate the Q-point on the dc load line.

Question: Was the Q-point for the drain-feedback biased E-MOSFET near the middle of the load line?

Step 14.   Following the procedure in Experiment 25, Step 6, change the IRF511 MOSFET
               transconductance coefficient (KP) from 0.042 to 0.02. Run the simulation and record the
               values of drain current (I_D) and drain-source voltage (V_{DS}).

               I_D = _____ V_{DS} = _____

NOTE: If you are using this manual in a lab environment, you should change the IRF511 MOSFET
instead of changing the MOSFET parameters.

Step 15.   Based on the values measured in Step 14, locate the new Q-point on the load line plotted in
               Step 12.

Question: Which bias circuit had the best Q-point stability, the E-MOSFET voltage-divider bias or the E-
MOSFET drain-feedback bias? **Explain.**

Step 16.   Return the IRF511 MOSFET transconductance coefficient (KP) back to 0.042.

NOTE: If you are using this manual in a lab environment, you should change back to the original IRF511 MOSFET.

Troubleshooting Problems

1.   Open circuit file FIG26-4 and run the simulation. Based on the measured voltages and currents, what is wrong with the 2N3796 D-MOSFET?

2.   Open circuit file FIG26-5 and run the simulation. Determine which circuit component is defective and state the defect (open or short) based on the measured voltages and currents.

        Defective component: _____   Defect: _____

3.   Open circuit file FIG26-6 and run the simulation. Determine which circuit component is defective and state the defect (open or short) based on the measured voltages and currents.

        Defective component: _____   Defect: _____

27

Name_____

Date_____

JFET Small-Signal
Common-Source Amplifier

Objectives:

1. Measure the voltage gain of a JFET small-signal common-source amplifier and compare the measured gain with the calculated gain.
2. Determine the phase difference between the input and output waveshapes for a one-stage JFET common-source amplifier.
3. Determine the effect of load resistance on the overall voltage gain.
4. Determine the output resistance of a JFET common-source amplifier.
5. Observe the effect of coupling capacitance on the dc offset of the ac output.
6. Determine the effect of unbypassed source resistance on the voltage gain of a JFET common-source amplifier.

Materials:

One MPF102 *n*-channel JFET
Capacitors: one 1 μF, one 470 μF
One 20 V dc power supply
One dual-trace oscilloscope
One function generator
Resistors: One 100 Ω, two 1 kΩ, two 1 MΩ

Theory:

Experiments 23 and 25 should be completed before attempting this experiment.

The purpose of **biasing** a JFET amplifier circuit is to establish an **operating point (Q-point)** in the active region of the characteristic curves to cause **linear** variations in drain current and drain-source voltage in response to ac input voltage variations. In applications where the ac input signal voltage variations are low, the drain current and drain-source voltage variations are relatively small. Therefore, the location of the operating point (Q-point) is not as critical as it would be for large input signal voltage variations, where it must be near the center of the load line. Amplifiers designed to handle low input voltage variations are called **small-signal amplifiers**. Because of their very **high input resistance**, FETs are often preferred over bipolar transistors for small-signal amplifier applications.

The amplifier in Figure 27-1 is a **self-biased** *n*-channel **JFET common-source small-signal amplifier**. Because of the extremely low gate current, the voltage across the gate resistor (R_G) is approximately 0 V dc, and the large resistance of R_G prevents loading of the ac signal source. Because the dc gate voltage is zero, an input coupling capacitor is not needed. Capacitor C_d is an **output coupling capacitor** that will prevent the dc drain voltage from being affected by the load resistance (R_L). This capacitor value needs to be large enough to have a very low reactance (X_C) at the lowest ac signal frequency expected. The dc voltage drop across the source resistor (R_S) causes the dc gate-source bias voltage (V_{GS}) to be negative. Capacitor C_S is a **bypass capacitor** that should have a much lower reactance (X_C) than the resistance of the source resistor (R_S) at the lowest ac signal frequency expected. This will cause the JFET source to be close to ground potential for the ac signal, but will still maintain the dc bias voltage on the JFET source. Without the JFET source bypass capacitor (C_S), the amplifier voltage gain (A_V) would be reduced.

In a JFET common-source amplifier, the ac signal causes the dc gate-source voltage to vary above and below the dc bias level, resulting in a drain current variation. This variation in drain current will cause the operating point (Q-point) to move up and down the **ac load line** and cause a large variation in drain current and drain voltage. When the input voltage increases, the drain (output) voltage decreases and when the input voltage decreases, the drain (output) voltage increases This causes the output to be **180 degrees out-of-phase** with the input for a **one-stage amplifier**. The ratio between the large drain (output) voltage variation and the small input voltage variation is the amplifier **voltage gain (A_V)**. Based on the ac peak-to-peak output and input voltages, the voltage gain of an amplifier is determined by dividing the ac peak-to-peak output voltage (V_o) by the ac peak-to-peak input voltage (V_{in}). Therefore,

$$A_V = \frac{V_o}{V_{in}}$$

The **expected voltage gain (A_V)** of the one-stage common-source JFET amplifier in Figure 27-1 is calculated by multiplying the **ac drain resistance (R_d)** by the **JFET transconductance (g_m)**. The ac drain resistance is equal to the parallel equivalent of the drain resistor (R_D) and the load resistor (R_L). Therefore,

$$R_d = \frac{R_D R_L}{R_D + R_L}$$

and

$$A_V = g_m R_d$$

For a common-source JFET amplifier with **unbypassed source resistance (R_S)**, the **voltage gain (A_V)** is

$$A_V = \frac{g_m R_d}{1 + g_m R_S}$$

The amplifier **ac output resistance (R_o)** can be determined from the ac peak-to-peak open circuit output voltage (V_{oc}) and the ac peak-to-peak output voltage (V_o) for a given load resistance (R_L) using Thevenin's theorem and solving the following equation for R_o.

$$\frac{V_o}{V_{oc}} = \frac{R_L}{R_o + R_L}$$

NOTE: Assume a load resistance (R_L) of 1 MΩ or greater to be an open circuit.

Figure 27-1 JFET Common-Source Amplifier

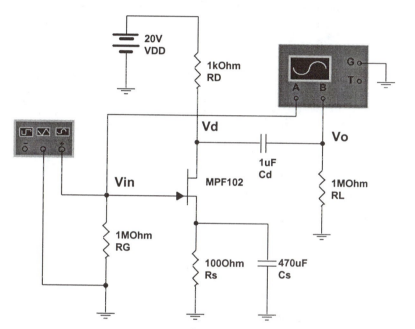

Procedure:

Step 1.   Open circuit file FIG27-1. Bring down the oscilloscope enlargement and make sure that the following settings are selected: Time base (Scale = 200 μs/Div, Xpos = 0, Y/T), Ch A (Scale = 50 mV/Div, Ypos = 0, DC), Ch B (Scale = 500 mV/Div, Ypos = 0, DC), Trigger (Pos edge, Level = 0, Sing, A). Bring down the function generator enlargement and make sure that the following settings are selected: *Sine Wave*, Freq = 1 kHz, Ampl = 100 mV, Offset = 0. Refer to the data collected in Experiments 23 and 25 and the JFET Self-Bias Circuit in Figure 25-1 for dc biasing currents and voltages (Q-point). **If Experiments 23 and 25 have not been completed, do those experiments before continuing.**

Step 2.   Run the simulation. Pause the simulation after one screen display on the oscilloscope. Record the ac peak-to-peak input voltage (V_{in}) and the ac peak-to-peak output voltage (V_o). Also record the phase difference between the input and output waveforms.

V_{in} = _____ V_o = _____

Phase difference = _____ degrees

Question: Were the input and output sine waves in-phase with each other? **Explain.**

Step 3.   Based on the voltage data in Step 2, determine the voltage gain (A_V) of the amplifier.

Step 4.   Using the value for the MPF102 transconductance (g_m) at the Q-point ($V_{GS} = -1$ V) determined in Step 9, Experiment 23, calculate the expected voltage gain (A_V) for the amplifier in Figure 27-1.

Question: How did the measured amplifier voltage gain compare with the calculated gain?

Step 5.   Change the load resistor (R_L) to 1 kΩ. Run the simulation and record the ac peak-to-peak input voltage (V_{in}) and the ac peak-to-peak output voltage (V_o). Adjust the oscilloscope settings as needed.

              V_{in} = _____ V_o = _____

Step 6.   Based on the voltage readings in Step 5, determine the new voltage gain (A_V).

Step 7.   Based on the value of the MPF102 transconductance (g_m) at the Q-point, determined in Step 9, Experiment 23, and the new value of the load resistance (R_L), calculate the new voltage gain (A_V).

Questions: How did the measured amplifier voltage gain compare with the calculated gain?

What effect did lowering the load resistance have on the overall voltage gain of the amplifier? **Explain.**

Step 8.   Based on the output voltage in Step 2 and the output voltage in Step 5, calculate the ac output resistance (R_o) of the amplifier. (Assume an R_L of 1 MΩ or greater to be an open circuit.)

Question: What was the relationship between the amplifier output resistance (R_o) and the value of the drain resistor (R_D)?

Step 9.   Change R_L back to 1 MΩ. Change the function generator amplitude to 500 mV, the oscilloscope Channel A input scale to 200 mV/Div, and the oscilloscope Channel B input scale to 5 V/Div.

Step 10.   Run the simulation. Record the dc offset voltage of the output (blue) waveshape. Move the Channel B oscilloscope wire to node V_d, run the simulation, and record the dc offset voltage of the output waveshape again.

DC offset at node V_o = _____

DC offset at node V_d = _____

Questions: What effect did the output coupling capacitor (C_d) have on the dc offset voltage of the ac output? **Explain.**

How did the offset voltage on the JFET drain (node V_d) compare with the dc drain voltage ($V_{DS} + V_S$) at the Q-point, measured in Experiment 25, Step 1?

Step 11.   Return the Channel B oscilloscope wire to node V_o. Set the Channel B scale input on the oscilloscope to 1 V/Div. Remove the source bypass capacitor (C_S) from the circuit. Run the simulation and record the ac peak-to-peak input voltage (V_{in}) and the ac peak-to-peak output voltage (V_o).

V_{in} = _____ V_o = _____

Step 12.   Based on the voltage readings in Step 11, determine the new voltage gain with unbypassed source resistance.

Step 13.   Based on the value of the MPF102 transconductance (g_m) at the Q-point, determined in Step 9, Experiment 23, and the value of the unbypassed source resistance (R_S), calculate the expected voltage gain (A_V) of the amplifier with the bypass capacitor removed.

Questions: How did the measured amplifier voltage gain compare with the calculated gain?

What effect did removing the bypass capacitor have on the overall amplifier voltage gain? **Explain.**

Troubleshooting Problems

1.   Open circuit file FIG27-2 and run the simulation. Locate the defective component and state the defect (short or open). You can use any instrument available and make any measurement desired.

Defective component: _____   Defect: _____

2.   Open circuit file FIG27-3 and run the simulation. Locate the defective component and state the defect (short or open). You can use any instrument available and make any measurement desired.

Defective component: _____   Defect: _____

3.   Open circuit file FIG27-4 and run the simulation. Locate the defective component and state the defect (short or open). You can use any instrument available and make any measurement desired.

        Defective component: _____   Defect: _____

4.   Open circuit file FIG27-5 and run the simulation. Locate the defective component and state the defect (short or open). You can use any instrument available and make any measurement desired.

        Defective component: _____   Defect: _____

5.   Open circuit file FIG27-6 and run the simulation. Locate the defective component and state the defect (short or open). You can use any instrument available and make any measurement desired.

        Defective component: _____   Defect: _____

28

Name_____

Date_____

MOSFET Small-Signal Common Source Amplifier

Objectives:

1. Measure the voltage gain of a MOSFET small-signal common-source amplifier and compare the measured gain with the calculated gain.
2. Determine the phase difference between the input and output waveshapes for a one-stage MOSFET common-source amplifier.
3. Determine the effect of load resistance on the overall voltage gain.
4. Determine the output resistance of a MOSFET common-source amplifier.
5. Observe the effect of coupling capacitance on the dc offset of the ac signal.

Materials:

One 2N3796 *n*-channel D-MOSFET
One 1 µF capacitor
One 20 V dc power supply
One dual-trace oscilloscope
One function generator
Resistors: two 4 kΩ, two 1 MΩ

Theory:

Experiments 24 and 26 should be completed before attempting this experiment.

The amplifier in Figure 28-1 is a **zero-biased** *n*-channel **D-MOSFET common-source small-signal amplifier** with the same bias circuit shown in Figure 26-1. Because of the extremely low gate current, the voltage across the gate resistor (R_G) is approximately 0 V. Because the MOSFET source is connected to ground, the **gate-source voltage (V_{GS})** is also 0 V. The large value of R_G causes the amplifier to have a high input impedance, preventing loading of the ac signal source. Because the dc gate voltage is zero, an input coupling capacitor is not needed. Capacitor C_d is an **output coupling capacitor** that will prevent the dc drain voltage from being affected by the load resistance (R_L). This capacitor value needs to be large enough to have a very low reactance (X_C) at the lowest ac signal frequency expected. The amplifier in Figure 28-1 has a very simple biasing arrangement because the D-MOSFET can operate in both the **depletion** and **enhancement modes** and be biased at $V_{GS} = 0$ V.

In a MOSFET common-source amplifier, the ac signal causes the dc gate-source voltage to vary above and below the dc bias level, resulting in a drain current variation. This variation in drain current will cause the **operating point (Q-point)** to move up and down the ac load line and cause a large variation in drain voltage. When the input voltage increases, the drain (output) voltage decreases and when the input voltage decreases, the drain (output) voltage increases This causes the output to be **180 degrees out-of-phase** with the input for a one-stage amplifier. The ratio between the large drain (output) voltage variation and the small input voltage variation is the **amplifier voltage gain (A_V)**. Based on the ac peak-to-peak output voltage (V_o) and the ac peak-to-peak input voltage (V_{in}), the voltage gain (A_V) of an amplifier is determined from

$$A_V = \frac{V_o}{V_{in}}$$

The **expected voltage gain (A_V)** of the one-stage D-MOSFET common-source amplifier in Figure 28-1 is calculated by multiplying the **ac drain resistance (R_d)** by the **MOSFET transconductance (g_m)**. The ac drain resistance is equal to the parallel equivalent of the drain resistor (R_D) and the load resistor (R_L). Therefore,

$$R_d = \frac{R_D R_L}{R_D + R_L}$$

and

$$A_V = g_m R_d$$

The amplifier **ac output resistance (R_o)** can be determined from the ac peak-to-peak open circuit output voltage (V_{oc}) and the ac peak-to-peak output voltage (V_o) for a given load resistance (R_L) using Thevenin's theorem and solving the following equation for R_o. Assume a load resistance (R_L) of 1 MΩ or greater to be an open circuit.

$$\frac{V_o}{V_{oc}} = \frac{R_L}{R_o + R_L}$$

Figure 28-1 D-MOSFET Common-Source Amplifier

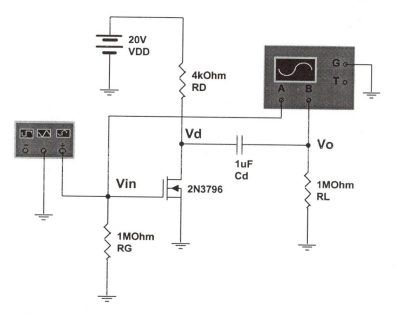

Procedure:

Step 1.   Open circuit file FIG28-1. Bring down the oscilloscope enlargement and make sure that the following settings are selected: Time base (Scale = 200 µs/Div, Xpos = 0, Y/T), Ch A (Scale = 50 mV/Div, Ypos = 0, DC), Ch B (Scale = 200 mV/Div, Ypos = 0, DC), Trigger (Pos edge, Level = 0, Sing, A). Bring down the function generator enlargement and make sure that the following settings are selected: *Sine Wave*, Freq = 1 kHz, Ampl = 100 mV, Offset = 0. Refer to the data collected in Experiments 24 and 26 and the D-MOSFET bias circuit in Figure 26-1 for dc biasing currents and voltages (Q-point). **If Experiments 24 and 26 have not been completed, do those experiments before continuing.**

Step 2.   Run the simulation. Pause the simulation after one screen display on the oscilloscope. Record the ac peak-to-peak input voltage (V_{in}) and the ac peak-to-peak output voltage (V_o). Also record the phase difference between the input and output waveforms.

V_{in} = _____ V_o = _____

Phase difference = _____ degrees

Question: Was the output in-phase with the input? **Explain.**

Step 3.   Based on the voltage data in Step 2, determine the voltage gain (A_V) of the amplifier.

Step 4.   Using the value for the 2N3796 transconductance (g_m) at the Q-point ($V_{GS} = 0$ V) determined in Step 8, Experiment 24, calculate the expected voltage gain (A_V) for the amplifier in Figure 28-1.

Question: How did the measured amplifier voltage gain compare with the calculated gain?

Step 5.   Change the load resistor (R_L) to 4 kΩ. Run the simulation and record the ac peak-to-peak input voltage (V_{in}) and the ac peak-to-peak output voltage (V_o). Adjust the oscilloscope settings as needed.

              V_{in} = _____ V_o = _____

Step 6.   Based on the voltage readings in Step 5, determine the new voltage gain (A_V).

Step 7.   Based on the value of the 2N3796 transconductance (g_m) at the Q-point, determined in Step 8, Experiment 24, and the new value of the load resistance (R_L), calculate the new voltage gain (A_V).

Questions: How did the measured amplifier voltage gain compare with the calculated gain?

What effect did lowering the load resistance have on the overall voltage gain of the amplifier? **Explain.**

Step 8.   Based on the output voltage in Step 2 and the output voltage in Step 5, determine the ac output resistance (R_o) of the amplifier. (Assume an R_L of 1 MΩ or greater to be an open circuit.)

Question: What was the relationship between the amplifier output resistance (R_o) and the value of the drain resistor (R_D)?

Step 9.   Change R_L back to 1 MΩ. Change the function generator amplitude to 500 mV, the oscilloscope Channel A scale input to 200 mV/Div, and the oscilloscope Channel B scale input to 5 V/Div.

Step 10.   Run the simulation. Record the dc offset voltage of the output (blue) waveshape. Move the Channel B oscilloscope wire to node V_d, run the simulation, and record the dc offset voltage of the output waveshape again.

DC offset at node V_o = _____

DC offset at node V_d = _____

Questions: What effect did the output coupling capacitor have on the dc offset voltage of the ac output? **Explain.**

How did the offset voltage of the ac output on the MOSFET drain compare with the dc drain-source voltage at the Q-point, measured in Experiment 26, Step 1?

How does the simplicity of the biasing circuit for the D-MOSFET amplifier compare with the simplicity of the biasing circuit for other amplifiers? **Explain.**

Troubleshooting Problems

1.   Open circuit file FIG28-2 and run the simulation. Locate the defective component and state the defect (short or open). You can use any instrument available and make any measurement desired.

        Defective component: _____   Defect: _____

2.   Open circuit file FIG28-3 and run the simulation. Locate the defective component and state the defect (short or open). You can use any instrument available and make any measurement desired.

        Defective component: _____   Defect: _____

3.   Open circuit file FIG28-4 and run the simulation. Locate the defective component and state the defect (short or open). You can use any instrument available and make any measurement desired.

        Defective component: _____   Defect: _____

IV

Operational Amplifiers

The following seven experiments involve operational amplifiers. First, you will do a dc and ac analysis of a bipolar junction transistor differential amplifier to help understand the basic circuitry of an operational amplifier. Next, you will measure the characteristics of the op-amp that will be used in the experiments in Part IV. You will then analyze noninverting and inverting op-amp amplifiers, op-amp comparator circuits, an op-amp summing amplifier, and op-amp integrator and differentiator circuits.

The circuits for the experiments in Part IV can be found on the enclosed disk in the OPAMP subdirectory.

29

Name_____

Date_____

Differential Amplifiers

Objectives:

1. Calculate the dc tail current for a differential amplifier and compare the calculated value with the measured value.
2. Calculate the dc collector currents in the balanced transistors for a differential amplifier and compare the calculated and measured values.
3. Calculate the balanced transistor dc collector voltages and compare the calculated and measured values.
4. Calculate the differential voltage gain of a differential amplifier and compare the calculated value with the measured value.
5. Determine the phase relationship between the output sine wave and the input sign wave when the input is applied to each base of a differential amplifier.
6. Determine the peak-to-peak sine wave differential output voltage of a differential amplifier and compare it to the peak-to-peak sine wave single-ended output voltage.
7. Calculate the differential amplifier common-mode voltage gain and compare the calculated value with the measured value.
8. Determine the common-mode rejection ratio (CMRR) of a differential amplifier and explain what it means regarding noise rejection.

Materials:

Two 2N3904 bipolar *npn* transistors
Two 15 V dc voltage supplies
Three 0–10 mA dc milliammeters
Two 0–10 V dc voltmeters
One dual-trace oscilloscope
One function generator
Resistors: two 100 Ω, three 2 kΩ

Theory:

Operational amplifiers (op-amps) are **integrated circuit (IC) dc amplifiers** with **high input resistance, high voltage gain**, and **low output resistance**. Transistors, diodes, and resistors are the only practical components to use when building IC circuits. For this reason, the IC designer has to use direct coupling between amplifier stages and eliminate the emitter bypass capacitor when building an IC operational amplifier. The **differential amplifier** is ideal for this kind of application because it eliminates the need for coupling capacitors and an emitter bypass capacitor, which makes it possible to use it in a frequency range down to dc. Therefore, in order to understand operational amplifiers, it is useful to have a basic understanding of differential amplifiers.

A **basic differential amplifier** circuit is shown in Figure 29-1. It consists of two identical common-emitter amplifiers in parallel with a single common emitter resistor and the transistor characteristics and collector resistors **matched** for balance. This is not a problem when building components on an IC chip because it is easier to match the component characteristics on the same semiconductor material. Although there are two transistors, the overall circuit is considered to be one-stage with a **differential input** between the transistor bases and a **differential output** between the transistor collectors. Because the components are on the same IC chip, any drift in component characteristics would likely be the same for each component. Because of this balance between components, the differential amplifier will maintain a **stable dc operating point**.

A differential amplifier can be operated in two different input and output modes, differential input or output and single-ended input or output. In the **differential output mode**, the output is taken between the two collectors with the collector on the right (V_{C2}) considered to be the positive side of the output and the collector on the left (V_{C1}) the negative side. In the **single-ended output mode**, the output is taken from the collector on the right (V_{C2}). Because the ac voltages on the transistor collectors are 180 degrees out-of-phase, the ac output voltage level for the differential output is double the ac output voltage level for the single-ended output. Therefore, the voltage gain for the differential output is double the voltage gain for the single-ended output. In the **differential input mode**, the input is applied between the two bases. In the **single-ended input mode**, the input is applied to one of the bases and the other base is grounded. When V_{B1} is used as the input, the output will not be inverted. Therefore, V_{B1} is called the **noninverting input**. When V_{B2} is used as the input, the output will be inverted. Therefore, V_{B2} is called the **inverting input**.

The circuit in Figure 29-1 will be used for a dc analysis of the differential amplifier. We will assume identical collector resistor (R_C) characteristics and identical transistor characteristics. The dc current through the common emitter resistor (R_E) is called the **tail current (I_T)**. The differential amplifier dc tail current is calculated by first determining the voltage across the emitter resistor (R_E), and then dividing by the value of R_E. Assuming that the dc base current in each transistor is negligible, the voltage on each base (V_B) is approximately zero. Therefore,

$$V_E = V_B - V_{BE} \cong 0 - V_{BE} = -V_{BE}$$

$$I_T \cong \frac{V_E - (-V_{EE})}{R_E} = \frac{V_{EE} - V_{BE}}{R_E}$$

The **dc collector currents (I_{C1} and I_{C2})** are approximately equal to the **dc emitter currents (I_{E1} and I_{E2})**. The dc emitter currents will be equal to each other if the two transistors are in balance. The tail current is equal to the sum of the emitter currents; and assuming perfect balance,

$$I_{E1} = I_{E2} = \frac{I_T}{2}$$

and

$$I_{C1} = I_{C2} \cong \frac{I_T}{2}$$

Each dc collector voltage can be determined by calculating the voltage across each collector resistor and subtracting this value from the dc supply voltage (V_{CC}). Therefore,

$$V_{C1} = V_{CC} - I_{C1}R_{C1}$$

$$V_{C2} = V_{CC} - I_{C2}R_{C2}$$

The **dc collector voltages (V_{C1} and V_{C2})** will be equal if the two collector currents (I_{C1} and I_{C2}) are equal, and the two collector resistors (R_{C1} and R_{C2}) are equal.

The circuit in Figure 29-2 will be used for an **ac analysis** of the differential amplifier. **The differential voltage gain ($A_{V(d)}$)** of a differential amplifier with a **single-ended output** is found by measuring the ac peak-to-peak voltage on the collector (V_{c2}) and the ac peak-to-peak voltage between the transistor bases ($V_{b1} - V_{b2}$). The differential voltage gain is

$$A_{(V)d} = \frac{V_{c2}}{V_{b1} - V_{b2}}$$

Because base 2 is grounded through the 100 Ω resistor,

$$A_{V(d)} = \frac{V_{c2}}{V_{b1} - 0} = \frac{V_{c2}}{V_{b1}}$$

The **expected differential voltage gain ($A_{V(d)}$)** of a differential amplifier with a **single-ended output** can be calculated based on the value of the collector resistors (R_C) and the **transistor emitter resistance (r_e)**. The expected differential voltage gain is

$$A_{V(d)} = \frac{R_C}{2r_e}$$

where $A_{V(d)}$ is the voltage gain between collector 2 and base 1. The voltage gain would be **double** the above value for a **differential output**.

Because the ac output voltage is proportional to the differential input voltage ($V_o = A_V(V_{b1} - V_{b2})$) and unwanted signals, such as **noise**, normally appear with the same magnitude and polarity on both inputs at the same time, the unwanted noise output should theoretically be zero. This noise rejection feature is one of the biggest advantages of the differential amplifier. In a real differential amplifier, the unwanted noise signal is not exactly zero. The ac peak-to-peak output voltage (V_{c2}) divided by the ac peak-to-peak **common mode voltage** on the bases ($V_{b1} = V_{b2}$) is the differential **amplifier common-mode voltage gain (A_{CM})**. Therefore,

$$A_{CM} = \frac{V_{c2}}{V_{b1}} = \frac{V_{c2}}{V_{b2}}$$

where V_{b1} is equal to V_{b2}.

Normally, A_{CM} is less than 1 and ideally should be 0 for complete noise rejection. The **expected common-mode voltage gain** can be calculated from

$$A_{CM} = \frac{R_C}{2R_E}$$

where R_C is the collector resistor value and R_E is the emitter resistor value.

The **common-mode rejection ratio (CMRR)** is the ratio of the differential voltage gain ($A_{V(d)}$) divided by the common-mode voltage gain (A_{CM}). It is a measure of how well the amplifier rejects unwanted noise entering at the transistor bases. Therefore,

$$CMRR = \frac{A_{V(d)}}{A_{CM}}$$

The higher the CMRR, the better the amplifier **signal-to-noise ratio**. The common-mode rejection ratio is often expressed in **decibels (dB)** as

$$CMRR = 20 \log \frac{A_{V(d)}}{A_{CM}}$$

Figure 29-1 Differential Amplifier–DC Analysis

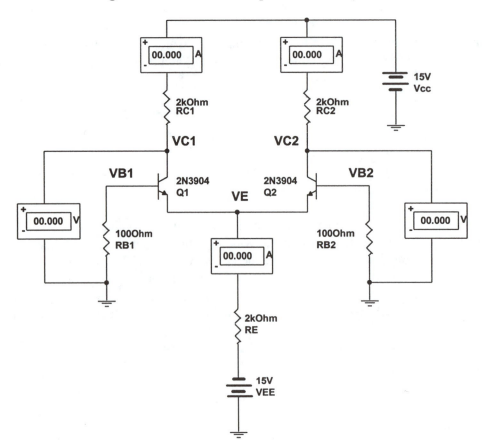

Figure 29-2 Differential Amplifier

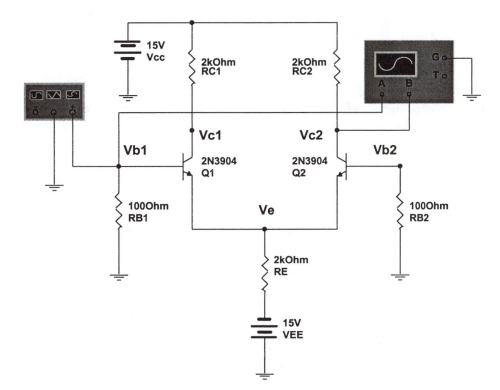

Procedure:

Step 1.   Open circuit file FIG29-1 and run the simulation. When steady state is reached, pause the simulation and record the dc tail current (I_T), the dc collector currents (I_{C1} and I_{C2}), and the dc collector voltages (V_{C1} and V_{C2}).

I_T = _____ I_{C1} = _____ I_{C2} = _____

V_{C1} = _____ V_{C2} = _____

Questions: What is the relationship between the dc tail current and the dc collector currents in the balanced transistors? **Explain.**

Are the dc collector currents equal? What does this tell you about the balance between the two transistors?

Are the dc collector voltages equal? What does this tell you about the balance between the two collector resistors?

Step 2.   Calculate the expected dc tail current (I_T) from the circuit component values. (Assume a value of 0.7 V for V_{BE}.)

Question: How did your calculated value for the dc tail current (I_T) compare with the measured value in Step 1?

Step 3.   Calculate the expected dc collector currents (I_{C1} and I_{C2}) for balanced transistors.

Question: How did your calculated values for the dc collector currents compare with the measured values?

Step 4.   Calculate the expected dc collector voltages (V_{C1} and V_{C2}) for balanced transistors.

Question: How did your calculated values for the dc collector voltages compare with the measured values?

Step 5.   Open circuit file FIG29-2. Bring down the oscilloscope enlargement and make sure that the following settings are selected: Time base (Scale = 500 μs/Div, Xpos = 0, Y/T), Ch A (Scale = 5 mV/Div, Ypos = 0, AC), Ch B (Scale = 500 mV/Div, Ypos = 0, AC), Trigger (Pos edge, Level = 0, Sing, A). Bring down the function generator enlargement and make sure that the following settings are selected: *Sine Wave*, Freq = 1 kHz, Ampl = 10 mV, Offset = 0. Run the simulation and record the ac peak-to-peak output voltage (blue) on collector 2 (V_{c2}) and the ac peak-to-peak input voltage (red) on base 1 (V_{b1}).

V_{c2} = _____ V_{b1} = _____

Step 6.   Based on the readings in Step 5, determine the differential voltage gain ($A_{V(d)}$) of the amplifier.

Step 7.   Based on the value of the circuit components and the value of the transistor ac emitter resistance (r_e) calculated in Experiment 11, Step 12, calculate the expected differential voltage gain ($A_{V(d)}$) between base 1 and collector 2.

Question: How did the calculated value for the differential voltage gain compare with the measured value? **Explain any difference.**

Step 8.   Record the phase difference between the input sine wave (V_{b1}) and the output sine wave (V_{c2}). Move the function generator output and the oscilloscope input A lead to base 2. Run the simulation again and record the phase difference between the new input sine wave (V_{b2}) and the output sine wave (V_{c2}).

Phase between V_{b1} and V_{c2} = _____ degrees

Phase between V_{b2} and V_{c2} = _____ degrees

Question: Based on the data collected in Step 8, which base is the inverting input and which base is the noninverting input for the differential amplifier circuit in Figure 29-2?

Step 9.   Move the oscilloscope ground lead to collector 1. Change the oscilloscope Channel B input
               to 1V/Div, and change the Channel A input from AC to 0. Change the oscilloscope trigger
               input to *Auto*. Run the simulation again and pause the simulation after one screen display on
               the oscilloscope. Record the ac peak-to-peak output voltage (V_o) between collectors.

               V_o = _____

Question: How did the ac peak-to-peak output voltage between collectors (differential output) compare
with the ac peak-to-peak single-ended output voltage between one collector and ground, in Step 5?
Explain.

Step 10.   Move the oscilloscope ground lead back to a ground terminal and connect both bases (V_{b1}
               and V_{b2}) together. Change the oscilloscope Channel A input back to AC and the Channel B
               input to 2mv/Div. Run the simulation until steady-state is reached, then pause the
               simulation. Record the ac peak-to-peak output voltage (V_{c2}) and input voltage (V_{b2}).

               V_{c2} = _____ V_{b2} = _____

Step 11.   Based on the voltages measured in Step 10, determine the common-mode voltage gain (A_{CM}).

Question: How did the common-mode voltage gain compare with the differential voltage gain? What
implication does this have regarding differential amplifier noise rejection?

Step 12.   Based on the circuit component values, calculate the expected common-mode voltage gain (A_{CM}).

Question: How did the calculated value for the common-mode voltage gain compare with the measured value?

Step 13.　　Based on the differential voltage gain ($A_{V(d)}$) measured in Steps 5 and 6 and the common-mode gain (A_{CM}) measured in Steps 10 and 11, calculate the common-mode rejection ratio (CMRR). Express your answer in decibels (dB).

Question: What is the meaning of common-mode rejection ratio (CMRR) for a differential amplifier? What is the advantage of a high CMRR?

Troubleshooting Problems

1.　　Open circuit file FIG29-3 and run the simulation. Locate the defective component and state the defect (short or open). You can use any instrument available and make any measurement desired.

　　　　Defective component: _____　　　　　　Defect: _____

2.　　Open circuit file FIG29-4 and run the simulation. Locate the defective component and state the defect (short or open). You can use any instrument available and make any measurement desired.

　　　　Defective component: _____　　　　　　Defect: _____

3.　　Open circuit file FIG29-5 and run the simulation. Locate the defective component and state the defect (short or open). You can use any instrument available and make any measurement desired.

　　　　Defective component: _____　　　　　　Defect: _____

4.   Open circuit file FIG29-6 and run the simulation. Locate the defective component and state the defect (short or open). You can use any instrument available and make any measurement desired.

Defective component:_____   Defect: _____

5.   Open circuit file FIG29-7 and run the simulation. Locate the defective component and state the defect (short or open). You can use any instrument available and make any measurement desired.

Defective component:_____   Defect:_____

Operational Amplifiers

Name_____

Date_____

Objectives:

1. Measure the inverting input current and the noninverting input current for an operational amplifier.
2. Calculate the input offset current and the input bias current using the measured values of the inverting and noninverting input currents.
3. Measure the output offset voltage and calculate the input offset voltage from the output offset voltage.
4. Measure the input resistance of an operational amplifier.
5. Measure the output resistance of an operational amplifier.
6. Measure the slew rate of an operational amplifier.

Materials:

One LM741 op-amp
Two 15 V dc voltage supplies
One 0–20 V dc voltmeter
Two 0–1 μA dc microammeters
One function generator
One dual-trace oscilloscope
Resistors: two 100 Ω, one 10 kΩ, one 100 kΩ
One variable 500 Ω potentiometer.

Theory:

An **operational amplifier (op-amp)** is an **integrated circuit (IC) dc amplifier** with **high input resistance**, **high voltage gain**, and **low output resistance**. Because it is a **multistage differential amplifier**, it has two input terminals, the **noninverting input (+)** and the **inverting input (–)**. Most op-amps have a class B push-pull emitter-follower single-ended output, although some op-amps have a differential output. With a positive and negative supply voltage, the single-ended output will be 0 V dc when the differential input is 0 V dc. With the proper use of **negative feedback**, the voltage gain, bandwidth, input impedance, and output impedance can be set to the desired values. In addition to being used as amplifiers, op-amps can be used to build oscillators, active filters, waveform converters, comparators, integrators and differentiators, and other interesting circuits.

There are three characteristics that are of interest when designing with op-amps. They are **the input bias**

current, the **input offset current**, and the **input offset voltage**. In an ideal op-amp with perfect balance between the components in the differential amplifiers, the two input currents would be equal, making the input offset current and voltage zero. In a real op-amp, there is not a perfect balance between differential amplifier components. Therefore, the input bias currents are not exactly equal and there is an input offset current and an input offset voltage.

The op-amp **input bias current (I_{BIAS})** is defined as the average value of the noninverting input current (I_1) and the inverting input current (I_2). The circuit for measuring the input bias current is shown in Figure 30-1. The input bias current can be calculated from

$$I_{BIAS} = \frac{I_1 + I_2}{2}$$

The op-amp **input offset current (I_{OS})** is defined as the absolute value of the difference between the two input currents. The input offset current indicates how closely the op-amp differential amplifier transistors are matched. If the transistors are identical, then the input offset current is zero. The circuit for measuring the input offset current is shown in Figure 30-1. The input offset current can be calculated from

$$I_{OS} = |I_1 - I_2|$$

An ideal op-amp produces zero dc output voltage when the input voltage is zero. In a real op-amp, a small dc voltage appears at the output when the input voltage is zero. The **input offset voltage (V_{OS})** is defined as the dc input voltage required to make the dc output voltage zero. Based on the measured **output offset voltage ($V_{O(off)}$)**, the input offset voltage (V_{OS}) is equal to the output offset voltage divided by the amplifier voltage gain. A circuit for measuring the output offset voltage ($V_{O(off)}$) is shown in Figure 30-2. The input offset voltage (V_{OS}) can be calculated from

$$V_{os} = \frac{V_{O(off)}}{A_V}$$

where the voltage gain of the amplifier in Figure 30-2 is calculated from

$$A_V = \frac{R_f}{R_1} + 1$$

The op-amp **differential input resistance** is the total resistance between the inverting and the noninverting inputs. The **common-mode input resistance** is the resistance between each input and ground. A circuit for measuring the op-amp common-mode input resistance is shown in Figure 30-3. The op-amp input resistance (R_{in}) is equal to the input voltage (V_{in}) divided by the input current (I_{in}). Therefore,

$$R_{in} = \frac{V_{in}}{I_{in}}$$

The **op-amp output resistance** can be measured by measuring the open circuit output voltage ($R_L \geq$

100 kΩ can be considered an open circuit output). Then keep reducing the load resistance (R_L) value until the output voltage is equal to one-half the open circuit voltage. Measure the new resistance value of R_L. Based on Thevenin's theorem, this new value of R_L is equal to the op-amp output resistance (R_o). A circuit for measuring the op-amp output resistance is shown in Figure 30-3.

The op-amp **slew rate** (S_R) is defined as the maximum rate of change of the output voltage (V_O) in response to a step input voltage. In order to determine the maximum rate of change of the output voltage, a wide pulse that is high enough in voltage to saturate the op-amp is applied to the input. The slope of the rising edge of the output is the slew rate (S_R). Therefore,

$$S_R = \frac{\Delta V_O}{\Delta t}$$

A circuit for measuring the slew rate is shown in Figure 30-4. An op-amp connected for unity-gain (with unity feedback) gives the worst-case (slowest) slew rate.

The maximum frequency without distortion (f_{max}), called the **power bandwidth**, is proportional to the slew rate and inversely proportional to the output sine wave peak voltage (V_p). Therefore,

$$f_{max} = \frac{S_R}{2\pi V_P}$$

Slew rate distortion is one of the major limitations on the large-signal performance of an op-amp.

Figure 30-1 Operational Amplifier Input Offset and Bias Currents

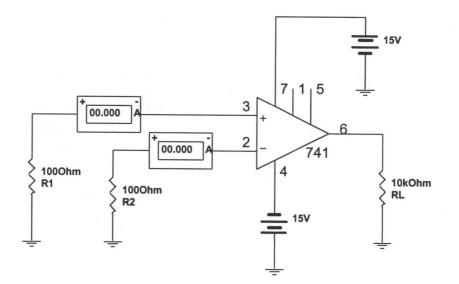

Figure 30-2 Operational Amplifier Input Offset Voltage

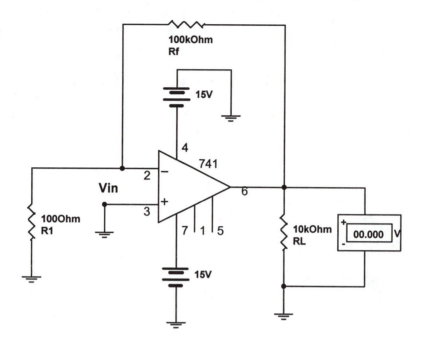

Figure 30-3 Operational Amplifier Input and Output Resistance

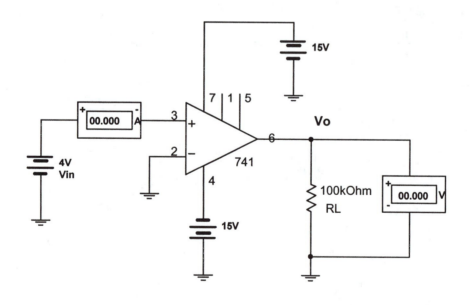

Figure 30-4 Operational Amplifier Slew Rate

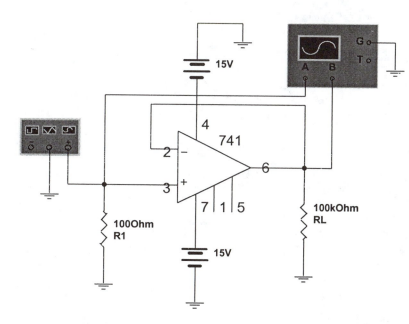

Procedure:

Step 1.   Open circuit file FIG30-1 and run the simulation. Record the input currents.

Noninverting input current (I_1) = _____

Inverting input current (I_2) = _____

Step 2.   Based on the input currents measured in Step 1, calculate the input bias current (I_{BIAS}) and the input offset current (I_{OS}).

Questions: How did your measured input bias current (I_{BIAS}) compare with the typical value for a 741 op-amp?

How did your measured input offset current (I_{OS}) compare with the typical value for a 741 op-amp?

Explain the difference between input offset current and input bias current.

Step 3. Open circuit file FIG30-2 and run the simulation. Record the output offset voltage ($V_{O(off)}$).

$$V_{O(off)} = \underline{\hspace{2cm}}$$

Step 4. Calculate the voltage gain (A_V) of the amplifier in Figure 30-2. Based on the output offset voltage measured in Step 3 and the voltage gain of the amplifier, calculate the input offset voltage (V_{OS}).

Question: How did your calculated input offset voltage (V_{OS}) compare with the typical value for a 741 op-amp?

Step 5. Open circuit file FIG30-3 and run the simulation. Record the input current (I_{in}) and the output voltage (V_o).

$$I_{in} = \underline{\hspace{2cm}} \qquad V_o = \underline{\hspace{2cm}}$$

Step 6. Keep reducing the value of load resistor R_L and run the simulation until the output voltage (V_o) is equal to one-half the value measured in Step 5. Record the new value of R_L. This is the output resistance (R_o) of the op-amp.

$$R_o = R_L = \underline{\hspace{2cm}}$$

Question: Was your measured op-amp output resistance high or low? Was this typical for a 741 op-amp?

Step 7.   Based on the value of I_{in} measured in Step 5 and the value of voltage source V_{in}, calculate the op-amp input resistance (R_{in}).

Question: Was your measured op-amp input resistance high or low? Was this typical for a 741 op-amp?

Step 8.   Open circuit file FIG30-4. Bring down the oscilloscope enlargement and make sure that the following settings are selected: Time base (Scale = 20 µs/Div, Xpos = 0, Y/T), Ch A (Scale = 5 V/Div, Ypos = 0, DC), Ch B (Scale = 5 V/Div, Ypos = 0, DC), Trigger (Pos edge, Level = 1 µV, Sing, A). Bring down the function generator enlargement and make sure that the following settings are selected: *Square Wave*, Freq = 5 kHz, Duty Cycle = 50%, Ampl = 5 V, Offset = 5 V. Run the simulation and measure the slope of the output curve (blue) to determine the slew rate (S_R) in V/µs. Record the value.

S_R = _____

Questions: How did your measured slew rate (S_R) compare with the typical value for a 741 op-amp?

What effect does slew rate have on the large-signal frequency response of an operational amplifier ?

31

Name_____

Date_____

Noninverting Voltage Amplifier

Objectives:

1. Calculate the expected voltage gain of an op-amp noninverting voltage amplifier and compare the calculated gain with the measured gain.
2. Determine the phase difference between the output and input sine waves for a single-stage noninverting voltage amplifier.
3. Calculate the expected dc output offset voltage based on the op-amp input offset voltage and compare the calculated value with the measured value.
4. Determine the significance of the dc output offset for different levels of signal inputs.

Materials:

One LM741 op-amp
Two 15 V dc voltage supplies
One function generator
One dual-trace oscilloscope
Resistors: one 1 kΩ, one 10 kΩ, two 100 kΩ

Theory:

The extremely high **open-loop voltage gain (A_{OL})** of an op-amp is too **unstable** to be of any use. A small noise voltage on the input will be amplified to such a high voltage that it will drive the op-amp into saturation. Also, unwanted oscillations can occur. If some of the amplifier output is fed back to the input 180° out-of-phase with the input (negative feedback), the new **closed-loop voltage gain (A_{CL})** will be lower and more stable than the open-loop voltage gain and independent of the magnitude of the open-loop voltage gain. The percentage of the op-amp output that is fed back to the input will determine the closed-loop voltage gain, the input resistance, the output resistance, and the bandwidth of the amplifier. The two basic amplifier configurations for wiring an op-amp with negative feedback are the **noninverting voltage amplifier** and the **inverting voltage amplifier**. In this experiment, you will study the noninverting voltage amplifier.

An op-amp wired as a **noninverting voltage amplifier** is shown in Figure 31-1. The input signal is applied to the op-amp **noninverting input (+)**. This will cause the output to be in-phase with the input. The output is fed back to the op-amp **inverting input (−)** through the feedback network of R_1 and R_2. The **feedback ratio**, $R_2 / (R_1 + R_2)$, determines the percentage of the output voltage fed back to the input. If the op-amp open-loop voltage gain (A_{OL}) is much greater than the desired closed-loop voltage gain, as is normally the case, the **expected closed-loop voltage gain (A_{CL})** of a noninverting voltage amplifier is practically equal to the inverse of the amplifier **feedback ratio (B)**. Therefore,

$$A_{CL} = \frac{1}{B} = \frac{R_1 + R_2}{R_2} = \frac{R_1}{R_2} + 1$$

This means that the closed-loop voltage gain (A_{CL}) is not dependent on the op-amp open-loop voltage gain (A_{OL}), but mostly depends on the value of the feedback resistors. This makes the closed-loop voltage gain very stable.

With **negative feedback**, the noninverting voltage amplifier **closed-loop input resistance** is approximately equal to the op-amp open-loop input resistance times the ratio of the open-loop voltage gain divided by the closed-loop voltage gain. For a 741 op-amp with A_{OL} = 200,000, R_{in} = 2 MΩ, and a closed loop voltage gain (A_{CL}) of 100, the closed-loop amplifier input resistance would be approximately 4000 MΩ. The **closed-loop output resistance** is approximately equal to the op-amp open-loop output resistance times the ratio of the closed-loop voltage gain divided by the open-loop voltage gain. For a 741 op-amp with A_{OL} = 200,000, R_o = 75 Ω, and a closed loop voltage gain (A_{CL}) of 100, the closed-loop amplifier output resistance would be approximately 0.04 Ω. This means that with negative feedback, the input resistance will be so high that it can be considered to be infinite (open) and the output resistance will be so low that it can be considered to be zero (short). These are ideal values for input resistance and output resistance.

Because of the small imbalance between transistors inside the op-amp, there will be some dc output voltage present when the input is zero. This amplifier **output offset voltage ($V_{O(off)}$)** will be equal to the op-amp **input offset voltage (V_{OS})**, measured in Experiment 30, multiplied by the amplifier closed-loop voltage gain (A_{CL}). Therefore,

$$V_{O(off)} = V_{OS} A_{CL}$$

Figure 31-1 Noninverting Voltage Amplifier

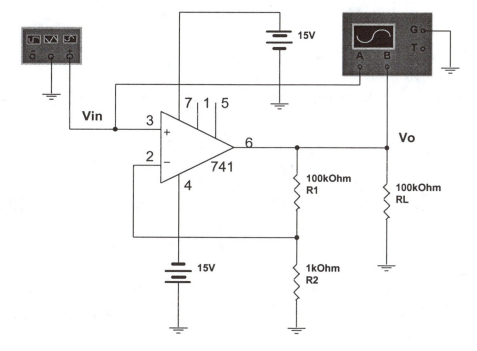

Procedure:

Step 1.   Open circuit file FIG31-1. Bring down the oscilloscope enlargement and make sure that the following settings are selected: Time base (Scale = 500 μs/Div, Xpos = 0, Y/T), Ch A (Scale = 5 mV/Div, Ypos = 0, DC), Ch B (Scale = 200 mV/Div, Ypos = 0, DC), Trigger (Pos edge, Level = 0, Sing, A). Bring down the function generator enlargement and make sure that the following settings are selected: *Sine Wave*, Freq = 1 kHz, Ampl = 5 mV, Offset = 0. Run the simulation and record the ac peak-to-peak input voltage (V_{in}) and the ac peak-to-peak output voltage (V_o). Also record the dc output offset voltage ($V_{O(off)}$) and the phase difference between the input and output sine waves.

V_{in} = _____ V_o = _____

$V_{O(off)}$ = _____   Phase difference = _____

Question: Were the output and input sine waves in-phase or out-of-phase? **Explain why.**

Step 2.   Based on the measured data in Step 1, determine the closed-loop voltage gain (A_{CL}) of the amplifier.

Step 3.   Based on the circuit component values, calculate the expected closed-loop voltage gain (A_{CL}).

Question: How did the calculated closed-loop voltage gain compare with the measured closed-loop voltage gain in Steps 1 and 2?

Step 4.   Based on the input offset voltage (V_{OS}) for the op-amp, calculated in Experiment 30, Step 4, calculate the dc output offset voltage ($V_{O(off)}$) expected for this amplifier gain.

Questions: How did the measured dc output offset voltage in Step 1 compare with the dc output offset voltage calculated from the input offset measured in Experiment 30?

What percentage of the peak output voltage in Step 1 was the dc output offset voltage?

Did the dc offset have much effect on the output? **Explain.**

Step 5.   Change the value of R_1 from 100 kΩ to 10 kΩ. Change the function generator amplitude to 100 mV, oscilloscope Channel A input to 100 mV/Div, and oscilloscope Channel B input to 500 mV/Div. Run the simulation and record the ac peak-to-peak input voltage (V_{in}), the ac peak-to-peak output voltage (V_o), and the dc output offset voltage ($V_{O(off)}$).

$V_{in} = $ _____ $V_o = $ _____ $V_{O(off)} = $ _____

Step 6.   Based on the measured data in Step 5, determine the new closed-loop voltage gain (A_{CL}) of the amplifier.

Step 7.   Based on the circuit component values, calculate the expected closed-loop voltage gain (A_{CL}).

Questions: How did the calculated closed-loop voltage gain compare with the measured closed-loop voltage gain in Steps 5 and 6?

When R_1 was changed, what effect did it have on the closed-loop voltage gain? **Explain.**

Step 8.   Based on the input offset voltage (V_{OS}) for the op-amp, calculate the dc output offset voltage ($V_{O(off)}$) expected for this amplifier gain.

Questions: What percentage of the peak output voltage in Step 5 was the dc output offset voltage, calculated in Step 8? Does this explain why the dc output offset voltage in Step 5 appeared to be zero?

What is the relationship between the peak input signal voltage and the dc input offset voltage in order to make the dc output offset voltage be negligible?

Troubleshooting Problems

1. Open circuit file FIG31-2 and run the simulation. Locate the defective component and state the defect (short or open). You can use any instrument available and make any measurement desired.

    Defective component: _____   Defect: _____

2. Open circuit file FIG31-3 and run the simulation. Locate the defective component and state the defect (short or open). You can use any instrument available and make any measurement desired.

    Defective component: _____   Defect: _____

3. Open circuit file FIG31-4 and run the simulation. Locate the defective component and state the defect (short or open). You can use any instrument available and make any measurement desired.

    Defective component: _____   Defect: _____

4. Open circuit file FIG31-5 and run the simulation. Locate the defective component and state the defect (short or open). You can use any instrument available and make any measurement desired.

    Defective component: _____   Defect: _____

32

Inverting Voltage Amplifier

Objectives:

1. Calculate the expected voltage gain of an op-amp inverting voltage amplifier and compare the calculated gain with the measured gain.
2. Determine the phase difference between the output and input sine waves for a single-stage inverting voltage amplifier.
3. Calculate the expected dc output offset voltage based on the op-amp input offset voltage and compare the calculated value with the measured value.
4. Determine the significance of the dc output offset for different levels of signal inputs.
5. Calculate the expected input resistance of an inverting voltage amplifier and compare the calculated resistance with the measured resistance.

Materials:

One LM741 op-amp
Two 15 V dc voltage supplies
One function generator
One dual-trace oscilloscope
Resistors: two 1 kΩ, one 10 kΩ, two 100 kΩ

Theory:

An op-amp wired as an **inverting voltage amplifier** is shown in Figure 32-1. The input signal is applied to the op-amp **inverting input (–)** through resistor R_1. This will cause the output to be 180 degrees out-of-phase with the input. The output is fed back to the op-amp inverting input (–) through the feedback network of R_1 and R_f. The feedback ratio, R_1 / R_f, determines the percentage of the output voltage fed back to the input. If the op-amp **open-loop voltage gain (A_{OL})** is much greater than the desired closed-loop voltage gain, as is normally the case, the **expected closed-loop voltage gain (A_{CL})** of an inverting voltage amplifier is practically equal to the inverse of the amplifier **feedback ratio (B)**. Therefore,

$$A_{CL} = -\frac{1}{B} = -\frac{R_f}{R_1}$$

This means that the **closed-loop voltage gain (A_{CL})** is not dependent on the op-amp open-loop voltage gain (A_{OL}), but only depends on the value of the feedback resistors. This makes the closed-loop voltage gain very **stable**. The negative voltage gain indicates **180 degrees phase inversion** between the input and the output.

With **negative feedback**, the inverting voltage amplifier **closed-loop input resistance (R_{in})** is approximately equal to the feedback resistance (R_f) divided by the open-loop voltage gain (A_{OL}) added to the resistance of R_1. Therefore,

$$R_{in} \cong R_1 + \frac{R_f}{A_{OL}}$$

For a 741 op-amp with an A_{OL} of 200,000 and a typical feedback resistance R_f of 100 kΩ, the closed-loop amplifier input resistance would be approximately equal to R_1. The amplifier input resistance can be measured by first determining the ac peak-to-peak voltage on node V_g. **The ac peak-to-peak input current (I_{in})** can be determined from

$$I_{in} = \frac{V_{in} - V_g}{R_1}$$

The input resistance (R_{in}) can be determined from

$$R_{in} = \frac{V_{in}}{I_{in}}$$

In an inverting amplifier circuit, the op-amp inverting input is practically at ground potential (virtual ground) because the voltage between the + and − inputs of the op-amp is so small and the current in the 1 kΩ resistor attached to the + input is so small. This places the left side of R_f near ground potential, putting it in parallel with the op-amp output resistance. Therefore, the **closed-loop output resistance** is approximately equal to the op-amp open-loop output resistance R_o in parallel with R_f. For a 741 op-amp with an R_o of 75 Ω, the closed-loop amplifier output resistance would be approximately equal to the op-amp open loop output resistance (R_o) for most values of R_f.

Because of the small imbalance between transistors inside the op-amp, there will be some dc output voltage present when the input is zero. This amplifier **output offset voltage ($V_{O(off)}$)** will be equal to the **op-amp input offset voltage (V_{OS})**, measured in Experiment 30, multiplied by the amplifier closed-loop voltage gain (A_{CL}). Therefore,

$$V_{O(off)} = V_{OS} A_{CL}$$

Figure 32-1 Inverting Voltage Amplifier

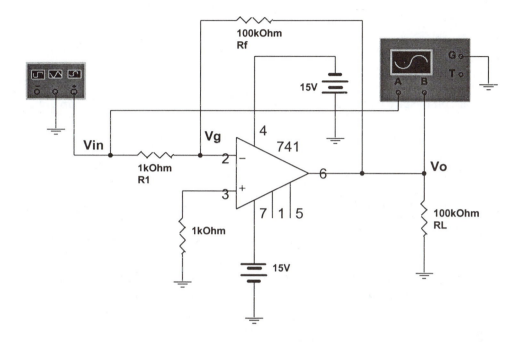

Procedure:

Step 1.   Open circuit file FIG32-1. Bring down the oscilloscope enlargement and make sure that the following settings are selected: Time base (Scale = 500 μs/Div, Xpos = 0, Y/T), Ch A (Scale = 5 mV/Div, Ypos = 0, DC), Ch B (Scale = 200 mV/Div, Ypos = 0, DC), Trigger (Pos edge, Level = 0, Sing, A). Bring down the function generator enlargement and make sure that the following settings are selected: *Sine Wave*, Freq = 1 kHz, Ampl = 5 mV, Offset = 0. Run the simulation and record the ac peak-to-peak input voltage (V_{in}) and the ac peak-to-peak output voltage (V_o). Also record the dc output offset voltage ($V_{O(off)}$) and the phase difference between the input and output sine waves.

V_{in} = _____ V_o = _____

$V_{O(off)}$ = _____   Phase difference = _____

Question: Was the output in-phase or out-of-phase with the input? **Explain why?**

Step 2.   Based on the measured data in Step 1, determine the closed-loop voltage gain (A_{CL}) of the amplifier.

Step 3.   Based on the circuit component values, calculate the expected closed-loop voltage gain (A_{CL}).

Question: How did the calculated closed-loop voltage gain compare with the measured closed-loop voltage gain in Steps 1 and 2?

Step 4.   Based on the input offset voltage (V_{OS}) for the op-amp, calculated in Experiment 30, Step 4, calculate the dc output offset voltage ($V_{O(off)}$) expected for this amplifier gain.

Questions: How did the measured dc output offset voltage in Step 1 compare with the calculated dc output offset voltage in Step 4?

What percentage of the peak output voltage in Step 1 was the dc output offset voltage?

Did the dc offset have much effect on the output? **Explain.**

Step 5.   Change the value of R_f from 100 kΩ to 10 kΩ. Change the function generator amplitude to 100 mV, oscilloscope Channel A input to 100 mV/Div, and oscilloscope Channel B input to 500 mV/Div. Run the simulation and record the ac peak-to-peak input voltage (V_{in}), the ac peak-to-peak output voltage (V_o), and the dc output offset voltage ($V_{O(off)}$).

V_{in} = _____ V_o = _____ $V_{O(off)}$ = _____

Step 6.   Based on the measured data in Step 5, determine the new closed-loop voltage gain (A_{CL}) of the amplifier.

Step 7.   Based on the circuit component values, calculate the expected closed-loop voltage gain (A_{CL}).

Questions: How did the calculated closed-loop voltage gain compare with the measured closed-loop voltage gain in Steps 5 and 6?

When R_f was changed, what effect did it have on the closed-loop voltage gain? **Explain.**

Step 8.   Based on the input offset voltage (V_{OS}) for the op-amp, calculate the dc output offset voltage ($V_{O(off)}$) expected for this amplifier gain.

Questions: What percentage of the peak output voltage in Step 5 was the dc output offset voltage, calculated in Step 8? Does this explain why the dc output offset voltage in Step 5 appeared to be zero?

What should be the relationship between the peak input signal voltage and the dc input offset voltage for an op-amp, in order to make the dc output offset voltage be negligible?

Step 9.   Move the Channel A oscilloscope lead to node V_g. Change the oscilloscope Channel A input to 1 mV/Div and change the trigger to *Auto*. Run the simulation. Pause the simulation and record the ac peak-to-peak voltage on node V_g.

                 V_g = _____

Step 10.   Based on the ac input voltage (V_{in}) measured in Step 5 and the ac voltage V_g, determine the input resistance (R_{in}) for the amplifier.

Step 11.   Based on the circuit component values, calculate the expected amplifier input resistance (R_{in}). (Assume an open loop op-amp voltage gain of A_{OL} = 200,000.)

Questions: How did the calculated input resistance in Step 11 compare with the value measured in Steps 5, 9, and 10?

How does the input resistance of the inverting amplifier compare with the input resistance of a noninverting amplifier?

Troubleshooting Problems

1. Open circuit file FIG32-2 and run the simulation. Locate the defective component and state the defect (short or open). You can use any instrument available and make any measurement desired.

    Defective component: _____   Defect: _____

2. Open circuit file FIG32-3 and run the simulation. Locate the defective component and state the defect (short or open). You can use any instrument available and make any measurement desired.

    Defective component: _____   Defect: _____

3. Open circuit file FIG32-4 and run the simulation. Locate the defective component and state the defect (short or open). You can use any instrument available and make any measurement desired.

    Defective component: _____   Defect: _____

33

Name_____

Date_____

Op-Amp Comparators

Objectives:

1. Examine the operation of the zero-level detection comparator.
2. Examine the operation of the inverting comparator.
3. Examine the operation of the nonzero-level detection comparator.
4. Examine the operation of the bounded comparator.
5. Examine the operation of the window comparator.
6. Examine a comparator with hysteresis, also known as a Schmitt trigger.

Materials:

Two LM 339 op-amp comparators
Two diodes
One 1N4733 5 V zener diode
Two 15 V dc voltage supplies
One 0–20 V dc voltage supply
Two 0–10 V dc voltmeters
One function generator
One dual-trace oscilloscope
Resistors (one each): 200 Ω, 300 Ω, 6 kΩ, 8 kΩ, 10 kΩ, 50 kΩ, 100 kΩ; two 2 kΩ

Theory:

When an operational amplifier is used to compare the amplitude of one voltage to another, it is called a **comparator**. When an op-amp is used as a comparator, it is used in the **open-loop** configuration with the input voltage on the op-amp noninverting input and a **reference voltage** on the op-amp inverting input. When the input voltage slightly exceeds the reference voltage, the op-amp is driven into **saturation** because of the high open-loop voltage gain, causing the output voltage to be up. When the input voltage is slightly less than the reference voltage, the op-amp is driven into negative saturation or cutoff, causing the output voltage to be down. The reference voltage is often referred to as the **trip point**. Therefore, for an op-amp comparator circuit, the trip point is the value of the input voltage that causes the output voltage to change state.

A **zero-level detection comparator** has a trip point (reference voltage) at the zero voltage level; a **positive reference voltage comparator** circuit has a trip point (reference voltage) at a positive voltage level; and a **negative reference voltage comparator** circuit has a trip point (reference voltage) at a negative voltage level. An **inverting comparator** has an output that goes down when the input is on the positive side of the trip point and goes up when the input is on the negative side of the trip point. An inverting comparator has the input voltage on the op-amp inverting input and the reference voltage on the op-amp noninverting input.

A zero-level detection comparator is shown in Figure 33-1. A zero-level detection inverting comparator is shown in Figure 33-2. The reference voltage for both comparator circuits is zero (ground potential).

Nonzero-level detection comparators are shown in Figures 33-3, 33-4, and 33-5. In Figures 33-3 and 33-4, the reference voltage is calculated from

$$V_{ref} = \frac{R_2}{R_1 + R_2} V$$

In Figure 33-5, the reference voltage is equal to the zener voltage of the 1N4733 diode.

In some applications, it is necessary to limit the output voltage levels to a value less than the saturation voltage (V_{sat}). A zener diode can be used as shown in Figure 33-6 to limit the output voltage swing to the zener voltage in one direction and the forward diode voltage in the other direction. This process is called **bounding**. Therefore, in a **bounded comparator**, the output voltage levels are limited to a particular voltage that is different from the saturation voltage (V_{sat}).

Two op-amp comparators wired as shown in Figure 33-7 form what is known as a **window comparator**. In a window comparator, the circuit detects when the input voltage is between two limits. When the input is between the limits, the output is low. When the input is not between the limits, the output is high. The upper and lower voltage limits are set by reference voltages V_a and V_b, as shown in Figure 33-7. The voltage divider rule can be used to calculate these voltage levels.

The **Schmitt trigger comparator**, shown in Figure 33-8, has **hysteresis**, which means that there is a higher trip point when the input voltage goes from low to high than when it goes from high to low. The Schmitt trigger comparator is less sensitive to input noise. The **upper trip point (V_{UTP})** can be calculated from

$$V_{UTP} = \frac{R_2}{R_1 + R_2} V_{sat}$$

The **lower trip point (V_{LTP})** can be calculated from

$$V_{LTP} = \frac{-R_2}{R_1 + R_2} V_{sat} = -V_{UTP}$$

where V_{sat} is the maximum output voltage.

Figure 33-1 Zero-Level Detection Comparator

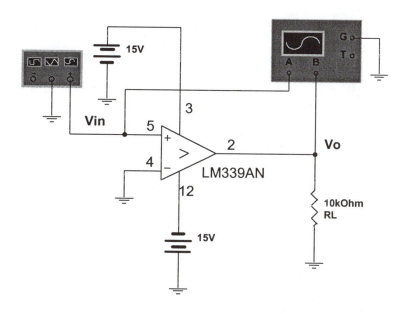

Figure 33-2 Zero-Level Detection Inverting Comparator

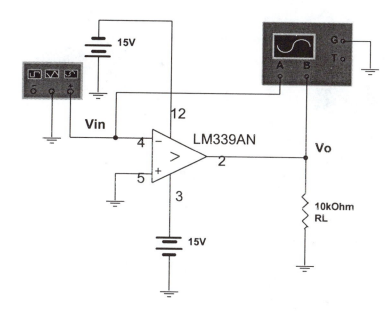

Figure 33-3 Nonzero-Level Detection Comparator
Positive Reference Voltage

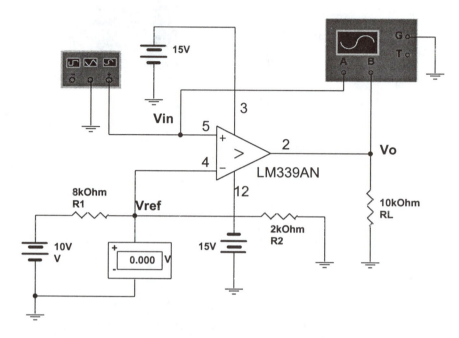

Figure 33-4 Nonzero-Level Detection Comparator
Negative Reference Voltage

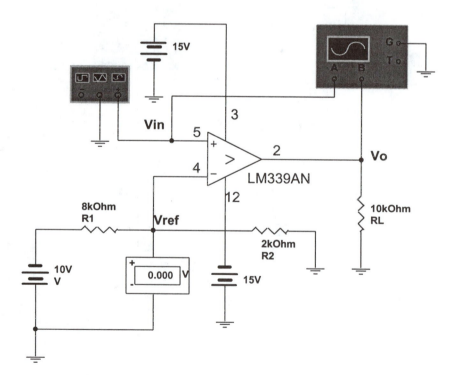

Figure 33-5 Nonzero-Level Detection Comparator
Zener Reference Voltage

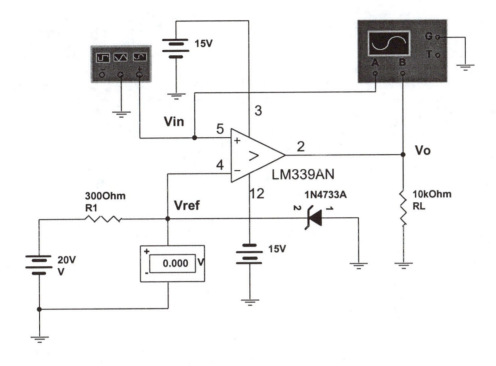

Figure 33-6 Bounded Comparator

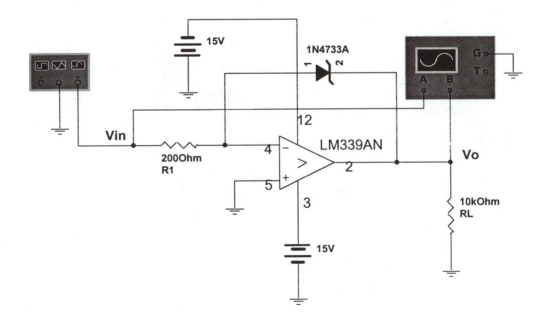

Figure 33-7 Window Comparator

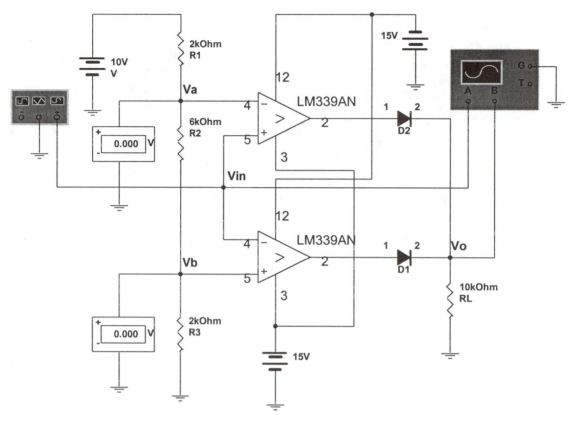

Figure 33-8 Comparator with Hysteresis–Schmitt Trigger

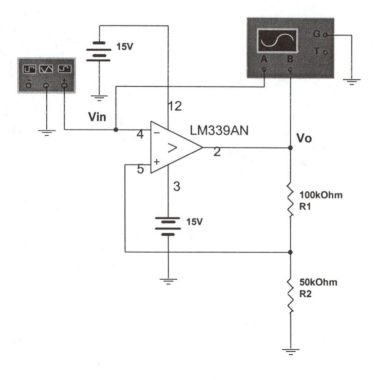

Procedure:

Step 1. Open circuit file FIG33-1. Bring down the oscilloscope enlargement and make sure that the following settings are selected: Time base (B/A), Ch A (Scale = 2 V/Div, Ypos = 0, DC), Ch B (Scale = 5 V/Div, Ypos = 0, DC), Trigger (Pos edge, Level = 1 mV, Sing, A). Bring down the function generator enlargement and make sure that the following settings are selected: *Sine Wave*, Freq = 100 Hz, Ampl = 5 V, Offset = 0. Run the simulation. You are plotting the output voltage (V_o) on the vertical axis (Channel B) and the input voltage (V_{in}) on the horizontal axis (Channel A). Draw the curve plot and *note the trip point on the curve plot.*

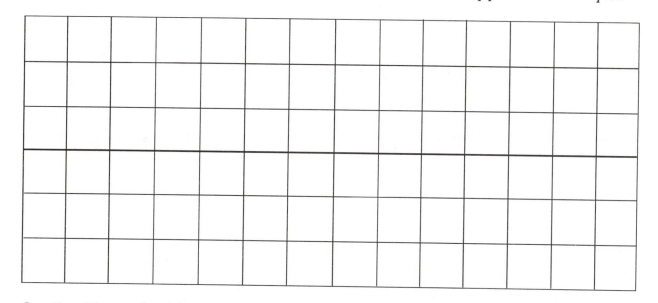

Question: The zero-level detection comparator trip point was at what input voltage level? **Explain why.**

Step 2. Click Y/T on the oscilloscope and make sure that the Time base scale is 2 ms/Div. Run the simulation again. You are plotting the input (red) voltage (V_{in}) and the output (blue) voltage (V_o) as a function of time. Draw the curve plots for V_{in} and V_o (on the next page). *Note the trip points on the V_{in} curve plot.*

Question: Can the zero-level detection comparator circuit produce a square wave output from a sine wave input? **Explain.**

Step 3.   Open circuit file FIG33-2. Make sure the instrument settings are the same as in Step 1 and run the simulation. You are plotting the output voltage (V_o) on the vertical axis (Channel B) and the input voltage (V_{in}) on the horizontal axis (Channel A). Draw the curve plot and *note the trip point on the curve plot.*

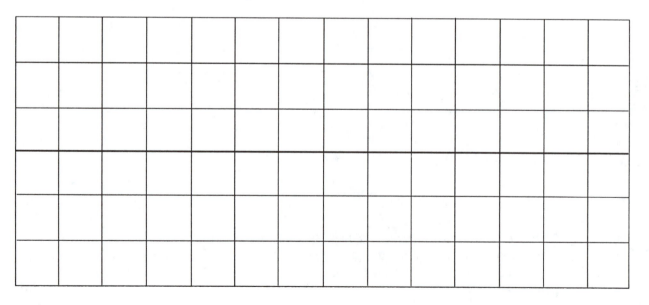

Step 4.   Click Y/T on the oscilloscope and make sure that the Time base scale is 2 ms/Div. Run the simulation again. You are plotting the input voltage (V_{in}) and the output voltage (V_o) as a function of time. Draw the curve plots for V_{in} and V_o. *Note the trip points on the V_{in} curve plot.*

Questions: What was the difference between the curve plots for the inverting comparator circuit and the noninverting comparator circuit? **Explain.**

What was the main difference between the noninverting comparator circuit in Figure 33-1 and the inverting comparator circuit in Figure 33-2?

Step 5.   Open circuit file FIG33-3. Make sure the instrument settings are the same as in Step 1 and run the simulation. You are plotting the output voltage (V_o) on the vertical axis (Channel B) and the input voltage (V_{in}) on the horizontal axis (Channel A). Draw the curve plot (on the following page) and *note the trip point on the curve plot.*

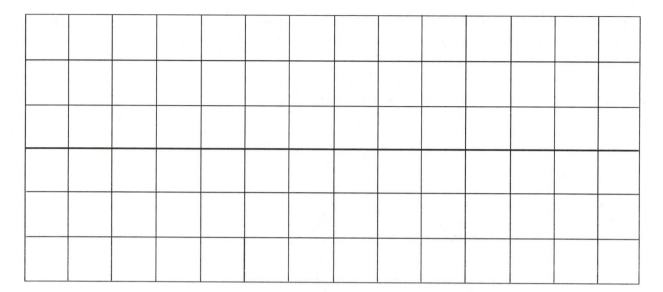

Step 6.   Click Y/T on the oscilloscope and make sure that the Time base scale is 2 ms/Div. Run the
            simulation again. You are plotting the input voltage (V_{in}) and the output voltage (V_o) as a
            function of time. Draw the curve plots for V_{in} and V_o. *Note the trip points on the V_{in} curve
            plot.* Also record the reference voltage (V_{ref}).

                    V_{ref} = _____

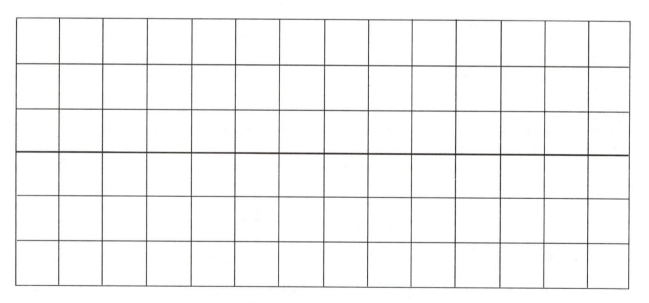

Step 7.   Calculate the expected reference voltage (V_{ref}) for the circuit in Figure 33-3 from the circuit
            component values.

Questions: How does the calculated value of V_{ref} compare with the measured value in Step 6?

How does the reference voltage (V_{ref}) in Step 6 compare with the trip point in Steps 5 and 6? **Explain.**

Step 8.   Open circuit file FIG33-4. Make sure the instrument settings are the same as in Step 1. Run the simulation. You are plotting the output voltage (V_o) on the vertical axis (Channel B) and the input voltage (V_{in}) on the horizontal axis (Channel A). Draw the curve plot and *note the trip point on the curve plot.*

Step 9.   Click Y/T on the oscilloscope and make sure that the Time base scale is 2 ms/Div. Run the simulation again. You are plotting the input voltage (V_{in}) and the output voltage (V_o) as a function of time. Draw the curve plots for V_{in} and V_o (on the following page). *Note the trip points on the V_{in} curve plot.* Also record the reference voltage (V_{ref}).

        V_{ref} = _____

Questions: Compare the results for the negative reference voltage comparator with the positive reference voltage comparator

What was the difference between the curve plots for the zero-level detection comparator, the positive reference voltage comparator, and the negative reference voltage comparator?

Step 10.   Open circuit file FIG33-5. Bring down the oscilloscope enlargement and make sure that the following settings are selected: Time base (B/A), Ch A (Scale = 5 V/Div, Ypos = 0, DC), Ch B (Scale = 5 V/Div, Ypos = 0, DC), Trigger (Pos edge, Level = 1 mV, Sing, A). Bring down the function generator enlargement and make sure that the following settings are selected: *SineWave*, Freq = 100 Hz, Ampl = 10 V, Offset = 0. Run the simulation. You are plotting the output voltage (V_o) on the vertical axis (Channel B) and the input voltage (V_{in}) on the horizontal axis (Channel A). Draw the curve plot (on the following page) and *note the trip point on the curve plot.*

Step 11.   Click Y/T on the oscilloscope and make sure that the Time base scale is 2 ms/Div. Run the
            simulation again. You are plotting the input voltage (V_{in}) and the output voltage (V_o) as a
            function of time. Draw the curve plots for V_{in} and V_o. *Note the trip points on the V_{in} curve
            plot.* Also record the reference voltage (V_{ref}).

            V_{ref} = _____

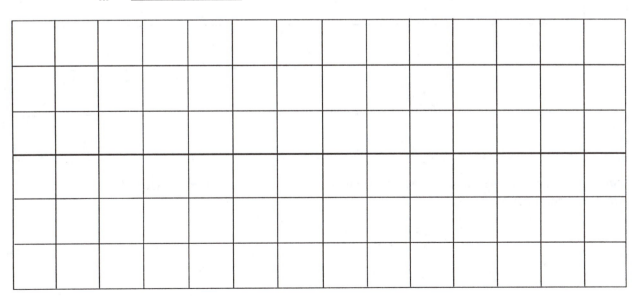

Questions: Compare the results for the zener reference voltage comparator with the positive reference voltage comparator.

How does the reference voltage (V_{ref}) compare with the zener diode voltage (V_z) from Experiment 3?

Step 12.   Open circuit file FIG33-6. Bring down the oscilloscope enlargement and make sure that the following settings are selected: Time base (Scale = 2 ms/Div, Xpos = 0, Y/T), Ch A (Scale = 1 V/Div, Ypos = 0, DC), Ch B (Scale = 2 V/Div, Ypos = 0, DC), Trigger (Pos edge, Level = 1 mV, Sing, A). Bring down the function generator enlargement and make sure that the following settings are selected: *Sine Wave*, Freq = 100 Hz, Ampl = 2 V, Offset = 0. Run the simulation. You are plotting the input voltage (V_{in}) and the output voltage (V_o) as a function of time. Draw the curve plots for V_{in} and V_o. *Note the trip points on the V_{in} curve plot. Also note the maximum and minimum output voltages on the curve plot.* Record the zener voltage (V_z) for the IN4733 determined from Experiment 3.

V_z = _____

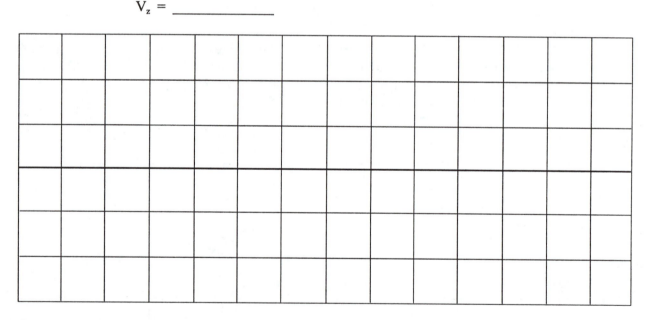

Question: What was the maximum output voltage for the bounded comparator? How does this compare with the zener diode voltage?

Step 13.   Remove the zener diode, reverse it, and then reconnect it. Repeat Step 12, draw the curve plot, and record the same results on the curve plot as in Step 12. Notice the difference between these results and the results in Step 12.

Question: What was the maximum output voltage for the bounded comparator? How does this compare with the curve plot in Step 12?

Step 14.   Open circuit file FIG33-7. Bring down the oscilloscope enlargement and make sure that the following settings are selected: Time base (Scale = 2 ms/Div, Xpos = 0, Y/T), Ch A (Scale = 5 V/Div, Ypos = 0, DC), Ch B (Scale = 5 V/Div, Ypos = 0, DC), Trigger (Pos edge, Level = 1 mV, Sing, A). Bring down the function generator enlargement and make sure that the following settings are selected: *Sine Wave*, Freq = 100 Hz, Ampl = 5 V, Offset = 5 V. Run the simulation. You are plotting the input voltage (V_{in}) and the output voltage (V_o) as a function of time. Draw the curve plots for V_{in} and V_o. *Note the trip points on the curve plots.* Record V_a and V_b.

V_a = _____  V_b = _____

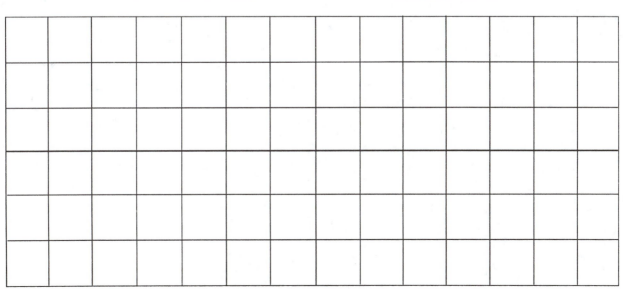

Question: What was the range of input voltages that produce a low output in the window comparator circuit?

Step 15.   Based on the circuit component values, calculate V_a and V_b in Figure 33-7 using the voltage divider rule.

Question: How did the calculated values for V_a and V_b compare with the measured values in Step 14?

Step 16.   Open circuit file FIG33-8. Bring down the oscilloscope enlargement and make sure that the following settings are selected: Time base (Scale = 2 ms/Div, Xpos = 0, Y/T), Ch A (Scale = 5 V/Div, Ypos = 0, DC), Ch B (Scale = 5 V/Div, Ypos = 0, DC), Trigger (Pos edge, Level = 0 V, Sing, A). Bring down the function generator enlargement and make sure that the following settings are selected: *Sine Wave*, Freq = 100 Hz, Ampl = 10 V, Offset = 0. Run the simulation. You are plotting the input voltage (V_{in}) and the output voltage (V_o) as a function of time for the Schmitt trigger circuit. Draw the curve plots for V_{in} and V_o. *Note the upper trip point (V_{UTP}) and the lower trip point (V_{LTP}) on the V_{in} curve plot.* Also note the maximum output voltage (V_{sat})

Step 17.   Based on the circuit component values and the value of V_{sat} determined in Step 16, calculate the expected upper trip point (V_{UTP}) and the expected lower trip point (V_{LTP}).

Question: How did the calculated values for the upper and lower trip points for the Schmitt trigger compare with the measured values in Step 16?

Step 18.   Click B/A on the oscilloscope and run the simulation. You are plotting the output voltage (V_o) on the vertical axis (Channel B) and the input voltage (V_{in}) on the horizontal axis (Channel A). Draw the curve plot. *Note the upper and lower trip points on the hysteresis loop.*

Question: What is the advantage of the different trip point voltages in the Schmitt trigger over the regular comparator circuit regarding input noise?

34

Name_____

Date_____

Op-Amp Summing Amplifier

Objectives:

1. Study the operation of the op-amp summing amplifier.
2. Analyze a summing amplifier with dc input voltages.
3. Analyze a summing amplifier with one ac input and one dc input.
4. Analyze a summing amplifier with all ac inputs.

Materials:

One LM741 op-amp
Four dc voltage supplies
Four 0–2 mA dc milliammeters
One 0–10 V dc voltmeter
One function generator
One dual-trace oscilloscope
Resistors: two 2.5 kΩ, three 5 kΩ, one 10 kΩ

Theory:

The **summing amplifier** is a variation of the **inverting amplifier**, except it has two or more inputs. In the summing amplifier, shown in Figure 34-1, the op-amp inverting input is considered to be **virtual ground** because of the very low voltage between the positive and negative op-amp inputs. Therefore, the input currents (I_1 and I_2) can be calculated from

$$I_1 = \frac{V_1}{R_1} \quad \text{and} \quad I_2 = \frac{V_2}{R_2}$$

From Kirchhoff's current law, the total current (I_t) can be determined from

$$I_t = I_1 + I_2$$

Because of the high op-amp input resistance, the **feedback current (I_f)** can be determined from

$$I_f \cong I_t = I_1 + I_2$$

Because the op-amp inverting input is at virtual ground potential, the summing amplifier **output voltage** (V_o) is approximately equal to the voltage across the feedback resistor (R_f) and is negative. Therefore,

$$V_o \cong -I_f R_f$$

Because $I_f = I_1 + I_2$, $I_1 = V_1/R_1$, and $I_2 = V_2/R_2$,

$$V_o \cong -(I_1 + I_2)R_f = -\left(\frac{V_1}{R_1} + \frac{V_2}{R_2}\right)(R_f) = -\left(\frac{R_f}{R}\right)(V_1 + V_2)$$

where $R = R_1 = R_2$. Therefore, the output voltage (V_o) is proportional to the sum of the input voltages ($V_1 + V_2$).

For the circuits in Figures 34-2 and 34-3, the output voltage (V_o) can be found from

$$V_o \cong -\left(\frac{V_1}{R_1} + \frac{V_2}{R_2}\right)(R_f) = -\left(\frac{R_f}{R}\right)(V_1 + V_2)$$

where $R = R_1 = R_2$.

Figure 34-1 Summing Amplifier with DC Input Voltages

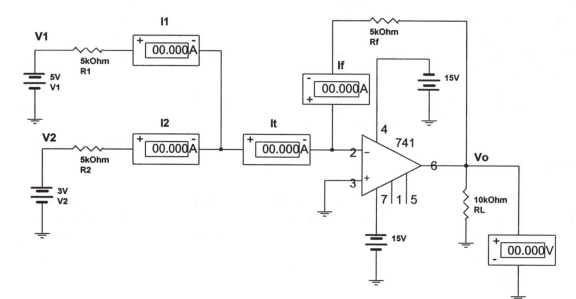

Figure 34-2 Summing Amplifier with One AC and One DC Input Voltage

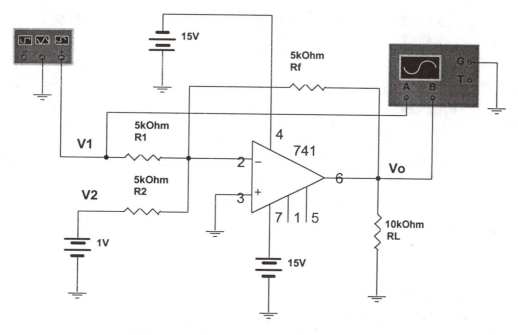

Figure 34-3 Summing Amplifier with AC Input Voltages

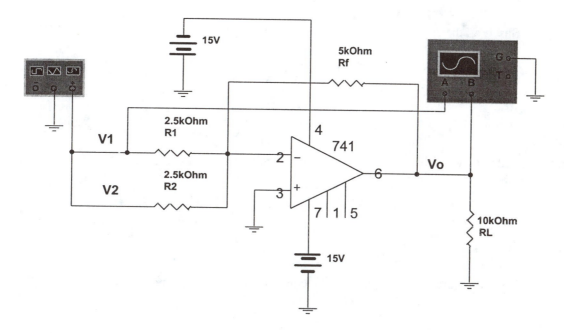

Procedure:

Step 1.   Open circuit file FIG34-1. Run the simulation and record I_1, I_2, I_t, I_f, V_1, V_2, and V_o.

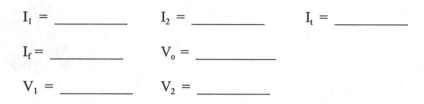

$I_1 = $ _____ $I_2 = $ _____ $I_t = $ _____

$I_f = $ _____ $V_o = $ _____

$V_1 = $ _____ $V_2 = $ _____

Step 2.   Based on the circuit component values, calculate the expected values for I_1, I_2, I_t, and I_f.

Question: How did your calculated values for I_1, I_2, I_t, and I_f compare with the measured values in Step 1?

Step 3.   Based on the current values calculated in Step 2, calculate the expected value of V_o.

Step 4.   Based on the values of V_1 and V_2, calculate the expected value of V_o.

Questions: How did your calculated value for V_o compare with the measured value in Step 1?   Why was it negative?

What was the relationship between the value of V_o and the input voltages (V_1 and V_2)?

How did the value of V_o calculated from V_1 and V_2 in Step 4 compare with the value of V_o calculated from the current values in Step 3?

Step 5.   Open circuit file FIG34-2. Bring down the oscilloscope enlargement and make sure that the following settings are selected: Time base (Scale = 200 μs/Div, Xpos = 0, Y/T), Ch A (Scale = 1 V/Div, Ypos = 0, DC), Ch B (Scale = 1 V/Div, Ypos = 0, DC), Trigger (Pos edge, Level = 0 V, Sing, A). Bring down the function generator enlargement and make sure that the following settings are selected: *Sine Wave*, Freq = 1 kHz, Ampl = 1 V, Offset = 0. Run the simulation and draw the input and output waveshapes. Record the output dc offset voltage.

Output dc offset = _____

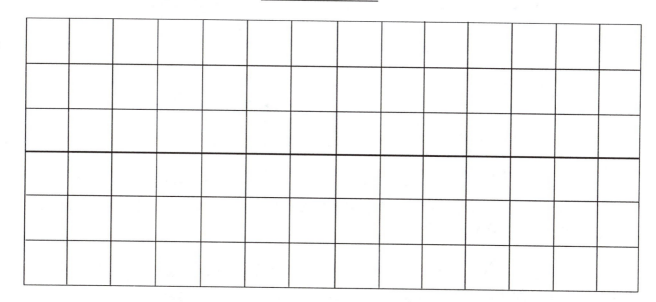

Step 6.   Based on the circuit component values, calculate the expected output voltage ($V_o(t)$) in terms of $V_1(t)$ and V_2.

Question: What was the relationship between $V_o(t)$ and the input voltages ($V_1(t)$ and V_2) in Steps 5 and 6? **Explain.**

Step 7.   Change R_2 to 2.5 kΩ. Run the simulation and draw the input and output waveshapes. Record the output dc offset voltage.

Output dc offset = _____

Step 8.   Based on the circuit component values, calculate the expected output voltage ($V_o(t)$) in terms of $V_1(t)$ and V_2.

Question: What was the relationship between $V_o(t)$ and the input voltages ($V_1(t)$ and V_2) in Steps 7 and 8? **Explain.**

Step 9. Open circuit file FIG34-3. Bring down the oscilloscope enlargement and make sure that the following settings are selected: Time base (Scale = 200 μs/Div, Xpos = 0, Y/T), Ch A (Scale = 1 V/Div, Ypos = 0, DC), Ch B (Scale = 2 V/Div, Ypos = 0, DC), Trigger (Pos edge, Level = 0 V, Sing, A). Bring down the function generator enlargement and make sure that the following settings are selected: *Sine Wave*, Freq = 1 kHz, Ampl = 1 V, Offset = 0. Run the simulation and draw the input and output waveshapes. *Note the input and output peak voltages on the curve plot.*

Step 10. Based on the circuit component values, calculate the expected output voltage ($V_o(t)$) in terms of $V_1(t)$ and $V_2(t)$.

Questions: What was the relationship between $V_o(t)$ and the input voltages ($V_1(t)$ and $V_2(t)$) in Steps 9 and 10? **Explain.**

Why are the output voltages negative for an op-amp summing amplifier?

Name_____

Date_____

Op-Amp Integrator and Differentiator

Objectives:

1. Demonstrate the operation of an op-amp integrator circuit.
2. Demonstrate the effect on the output of an integrator circuit when the input voltage is changed.
3. Demonstrate the effect on the output of an integrator circuit when the resistance or capacitance is changed.
4. Demonstrate the operation of an op-amp differentiator circuit.
5. Demonstrate the effect on the output of a differentiator circuit when the input slope is changed.
6. Demonstrate the effect on the output of a differentiator circuit when the resistance or capacitance is changed.
7. Demonstrate the purpose of the input resistor in a differentiator circuit.

Materials:

One LM741 op-amp
Two 15 V dc voltage supplies
One function generator
One dual-trace oscilloscope
Capacitors: 0.01 μF, 0.02 μF, 0.05 μF
Resistors: 500 Ω, 5 kΩ, 10 kΩ, 100 kΩ

Theory:

An **op-amp integrator** circuit, shown in Figure 35-1, simulates mathematical integration. The integral of a **step input** voltage produces an output **ramp function**. The slope of the output ramp ($\Delta V_o/\Delta t$) is proportional to the input step voltage (V_{in}). Therefore,

$$\frac{\Delta V_o}{\Delta t} = \frac{-V_{in}}{R_1 C}$$

for the circuit in Figure 35-1. The output is **inverted** because the input is applied to the op-amp inverting input. The **feedback resistor (R_f)** in Figure 35-1 prevents the op-amp from going into **saturation** by limiting the dc voltage gain of the op-amp ($A_V = R_f / R_1$), thus reducing the **output offset voltage**.

An **op-amp differentiator** circuit, shown in Figure 35-2, simulates mathematical differentiation, which is a process of determining the instantaneous **rate of change** of a function. A differentiator circuit produces an output voltage (V_o) that is proportional to the rate of change of the input voltage ($\Delta V_{in}/\Delta t$). Therefore,

$$V_o = -R_f\,C\!\left(\frac{\Delta V_{in}}{\Delta t}\right)$$

for the circuit in Figure 35-2. The output is **inverted** because the input is applied to the op-amp inverting input. The **input resistor (R_{in})** in Figure 35-2 prevents **oscillation** by limiting the closed-loop voltage gain at high frequencies.

Figure 35-1 Op-Amp Integrator

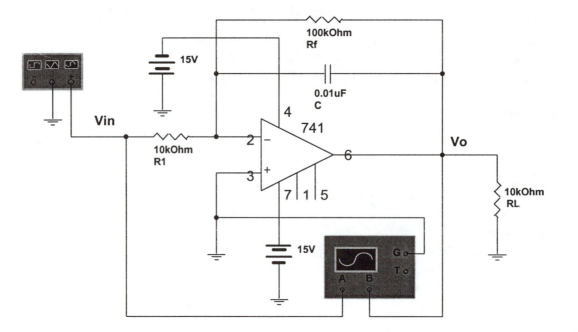

Figure 35-2 Op-Amp Differentiator

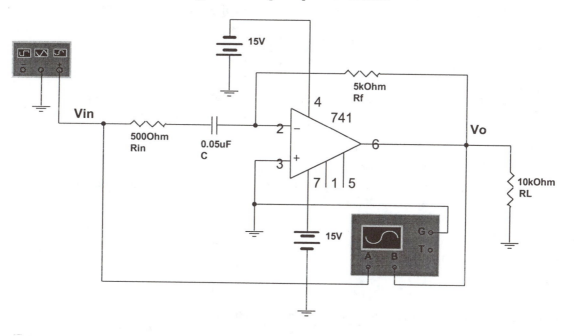

Procedure:

Step 1.   Open circuit file FIG35-1. Bring down the oscilloscope enlargement and make sure that the following settings are selected: Time base (Scale = 20 μs/Div, Xpos = 0, Y/T), Ch A (Scale = 5 V/Div, Ypos = 0, DC), Ch B (Scale = 5 V/Div, Ypos = 0, DC), Trigger (Pos edge, Level = 0 V, Sing, A). Bring down the function generator enlargement and make sure that the following settings are selected: *Square Wave*, Freq = 2 kHz, Duty Cycle = 50%, Ampl = 5 V, Offset = 0. Run the simulation and record the input (red) voltage (V_{in}) and the output (blue) slope in V/ms.

V_{in} = _____   Output slope = _____ V/ms

Question: Explain why the integrator output slope is inverted.

Step 2.   Based on the circuit component values and the value of the input voltage (V_{in}), calculate the expected output slope in V/ms.

Question: How did the calculated value of the integrator output slope compare with the measured value?

Step 3.   Change the function generator amplitude to 2 V and the oscilloscope Channel A and Channel B to 2 V/Div. Run the simulation and record the input voltage (V_{in}) and the output slope in V/ms.

V_{in} = _____   Output slope = _____ V/ms

Question: Was the integrator output slope dependent upon the value of V_{in}? **Explain.**

Step 4.   Based on the new value of the input voltage (V_{in}), calculate the expected output slope in V/ms.

Question: How did the calculated value of the integrator output slope compare with the measured value?

Step 5.   Change R_1 from 10 kΩ to 5 kΩ. Run the simulation and record the input voltage (V_{in}) and the new output slope in V/ms.

V_{in} = _____   Output slope = _____ V/ms

Question: Was the integrator output slope dependent upon the value of input resistor R_1? **Explain.**

Step 6.   Based on the new value of R_1, calculate the expected output slope in V/ms.

Question: How did the calculated value of the integrator output slope compare with the measured value?

Step 7.   Change Capacitor C to 0.02 µF. Run the simulation and record the input voltage (V_{in}) and the output slope in V/ms.

V_{in} = _____   Output slope = _____ V/ms

Question: Was the integrator output slope dependent upon the value of capacitor C? **Explain.**

Step 8.　　Based on the value of R_1 and the new value of capacitor C, calculate the expected output slope in V/ms.

Question: How did the calculated value of the integrator output slope compare with the measured value?

Step 9.　　Select *Nor* in the oscilloscope trigger section. Change R_1 back to 10 kΩ, C back to 0.01 μF, and the oscilloscope time base to 100 μs/Div. Run the simulation. After steady state is reached, pause the simulation. Draw the input and output waveshapes and label them.

Question: Was the output waveshape in Step 9 for the integrator circuit the integral of the input waveshape? **Explain.**

Step 10.   Open circuit file FIG35-2. Bring down the oscilloscope enlargement and make sure that the following settings are selected: Time base (Scale = 100 μs/Div, Xpos = 0, Y/T), Ch A (Scale = 5 V/Div, Ypos = 0, DC), Ch B (Scale = 5 V/Div, Ypos = 0, DC), Trigger (Pos edge, Level = 0 V, Sing, A). Bring down the function generator enlargement and make sure that the following settings are selected: *Ramp Function*, Freq = 1 kHz, Duty Cycle = 50%, Ampl = 5 V, Offset = 0. Run the simulation and draw the input and output waveshapes. Record the input slope in V/ms and the output pulse peak voltage (V_o).

Input slope = _____ V/ms V_o = _____

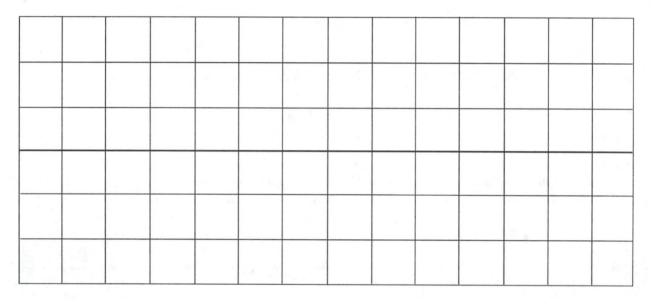

Questions: Was the output waveshape for the differentiator circuit in Step 10 the differential of the input waveshape? **Explain.**

Explain why the differentiator output pulse voltage is inverted.

Step 11.   Based on the circuit component values and the input slope in V/ms, calculate the expected output pulse peak voltage (V_o).

Question: How did the calculated value of the differentiator output pulse peak voltage (V_o) compare with the measured value?

Step 12.   Change the function generator frequency to 2 kHz and the oscilloscope time base to 50 µs/Div. Run the simulation and record the new input slope in V/ms and the output pulse peak voltage (V_o).

            Input slope = _____ V/ms V_o = _____

Question: Was the value of the differentiator output pulse peak voltage (V_o) dependent upon the input slope? **Explain.**

Step 13.   Calculate the expected output pulse peak voltage (V_o) for the new input slope.

Question: How did the calculated value of the differentiator output pulse peak voltage (V_o) compare with the measured value?

Step 14.   Return the frequency of the function generator to 1 kHz and the time base of the oscilloscope to 100 µs/Div. Change the feedback resistor (R_f) to 10 kΩ. Run the simulation and record the input slope in V/ms and the output pulse peak voltage (V_o).

            Input slope = _____ V/ms V_o = _____

Question: Was the differentiator output pulse peak voltage (V_o) dependent upon the value of the feedback resistor, R_f? **Explain.**

Step 15.   Based on the new value of R_f, calculate the expected output pulse peak voltage (V_o).

Question: How did the calculated value of the differentiator output pulse peak voltage (V_o) compare with the measured value?

Step 16.   Change capacitor C to 0.02 µF. Run the simulation and record the input slope in V/ms and the output pulse peak voltage (V_o).

Input slope = _____ V/ms V_o = _____

Question: Was the differentiator output voltage (V_o) dependent upon the value of capacitor C? **Explain.**

Step 17.   Based on the new value of C, calculate the expected output voltage (V_o).

Question: How did the calculated value of the differentiator output voltage (V_o) compare with the measured value?

Step 18.   Change capacitor C back to 0.05 µF. Remove resistor R_{in} and replace it with a short circuit. Run the simulation.

Question: What happened when the differentiator input resistor (R_{in}) was removed? **Explain why this happened.**

Troubleshooting Problems

1. Open circuit file FIG35-3 and run the simulation. Locate the defective component and state the defect (short or open). You can use any instrument available and make any measurement desired.

    Defective component: _____   Defect: _____

2. Open circuit file FIG35-4 and run the simulation. Locate the defective component and state the defect (short or open). You can use any instrument available and make any measurement desired.

    Defective component: _____   Defect: _____

3. Open circuit file FIG35-5 and run the simulation. Locate the defective component and state the defect (short or open). You can use any instrument available and make any measurement desired.

    Defective component: _____   Defect: _____

4. Open circuit file FIG35-6 and run the simulation. Locate the defective component and state the defect (short or open). You can use any instrument available and make any measurement desired.

    Defective component: _____   Defect: _____

5. Open circuit file FIG35-7 and run the simulation. Locate the defective component and state the defect (short or open). You can use any instrument available and make any measurement desired.

    Defective component: _____   Defect: _____

V

Amplifier Frequency Response

The following three experiments involve plotting amplifier frequency response curves and measuring the high and low cutoff frequencies. First, you will plot the low-frequency response curve and determine the low cutoff frequency for the small-signal CE transistor amplifier studied in Experiment 17. Next, you will plot the high-frequency response curve and determine the high cutoff frequency. Finally, you will plot the frequency response curves for an open-loop and a closed-loop op-amp amplifier and determine the effect of closed-loop amplifier gain on bandwidth.

The circuits for the experiments in Part V can be found on the enclosed disk in the FREQ subdirectory.

36

Low-Frequency Amplifier Response

Objectives:

1. Plot the amplifier low-frequency response curve for a common-emitter transistor amplifier.
2. Determine the midfrequency voltage gain from the dB gain for a common-emitter transistor amplifier.
3. Determine the low cutoff frequency for a common-emitter transistor amplifier.
4. Determine the low-frequency voltage gain roll-off in decibels (dB) per decade for a common-emitter transistor amplifier.
5. Determine the phase shift between the input and output waveshapes at the cutoff frequency and at midfrequency for a common-emitter transistor amplifier.
6. Calculate the input coupling network, output coupling network, and emitter-bypass network cutoff frequencies for a common-emitter transistor amplifier.
7. Select the dominant network in order to determine the expected low cutoff frequency, and compare the expected value with the measured value.
8. Determine the effect of decreasing the emitter-bypass capacitance on the amplifier low cutoff frequency.

Materials:

One 2N3904 bipolar transistor
Capacitors: one 1.0 μF, one 10 μF, one 20 μF, one 470 μF
One 20 V dc power supply
One dual-trace oscilloscope
One function generator
Resistors: one 660 Ω, two 2 kΩ, one 10 kΩ, one 200 kΩ

Theory:

The **decibel (dB)** is a common way of representing gain and is often used in plotting **amplifier gain** as a function of frequency. The basis for the decibel unit is derived from the **logarithmic response** of the human ear to the intensity of sound. The decibel is a measurement of the ratio of two power levels (power gain) or two voltage levels (voltage gain). **Voltage gain** is expressed in decibels (dB) from the actual voltage gain (A_V) as follows:

$$A_{dB} = 20 \log A_V$$

If A_V is greater than one, the dB gain is positive. If A_V is equal to one, the dB gain is zero. If A_V is less than one, the dB gain is negative and is referred to as attenuation.

The **cutoff frequency (f_C)** on a **gain-frequency plot** is the frequency where the voltage gain has dropped by **3 dB (0.707)** from the dB gain at midfrequency. At this frequency (f_C), the output power is one-half its midfrequency value. The difference between the high and low cutoff frequencies determines the **bandwidth** of an amplifier.

In this experiment, you will examine how amplifier **voltage gain** and **phase shift** are affected when the signal frequency approaches the amplifier **low cutoff frequency**. You will use the Bode plotter to study the low-frequency amplifier response of the **small-signal common-emitter amplifier** in Experiment 17. The amplifier circuit in Figure 17-1, Experiment 17, has been repeated in Figures 36-1 and 36-2.

Because capacitive reactance increases with decreasing frequency, amplifier **coupling and bypass capacitors** can no longer be considered short circuits at low frequencies. In Figure 36-1, **input coupling capacitor C_1** and the amplifier **input resistance (R_{in})** form an **input high-pass filter** with a **cutoff frequency (f_{in})** that can be calculated from

$$f_{in} = \frac{1}{2\pi R_{in} C_1}$$

where R_{in} can be found from Experiment 17, Steps 8 and 9.

In Figure 36-1, **output coupling capacitor C_2** and **output resistances R_C and R_L** form an **output high-pass filter** with a **cutoff frequency (f_o)** that can be calculated from

$$f_o = \frac{1}{2\pi(R_C + R_L)C_2}$$

In Figure 36-1, if you assume signal source resistance $R_S \cong 0$, **emitter bypass capacitor C_3** and the parallel combination of **emitter resistance r_e** and **emitter resistor R_E** form a **high-pass filter**. If you assume transistor emitter resistance $r_e \langle\langle R_E$, the cutoff frequency (f_e) can be calculated from

$$f_e = \frac{1}{2\pi r_e C_3}$$

where r_e is the transistor emitter resistance measured in Experiment 11, Step 12.

The voltage gain will drop **20 dB per decade** of frequency reduction below the cutoff frequency for each high-pass filter network. The **phase difference** between the input and the output will shift by **45 degrees** at the cutoff frequency from what it was at midfrequency.

The low cutoff frequency for the amplifier is determined by establishing which of the three high-pass filters has the highest cutoff frequency. This network is called the **dominant network** and its cutoff frqeuency determines the low cutoff frequency for the amplifier (voltage gain 3 dB down from midfrequency gain).

Figure 36-1 Common-Emitter Amplifier Low-Frequency Response

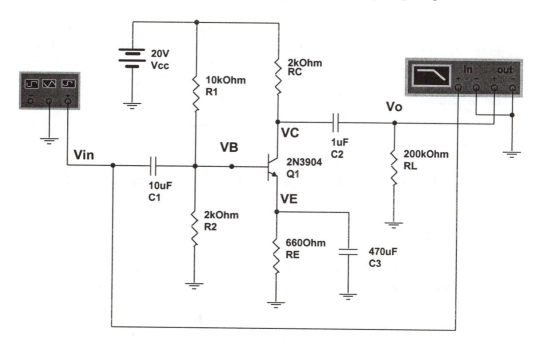

Figure 36-2 Common-Emitter Amplifier Low-Frequency Response

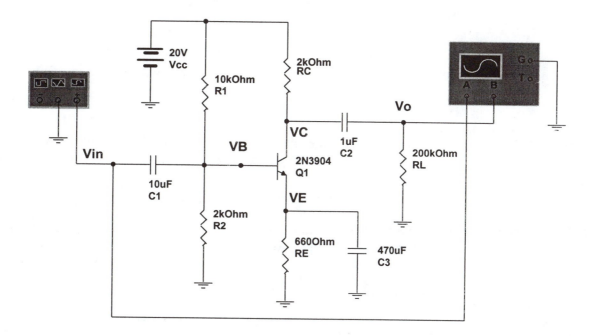

Procedure:

Step 1.   Open circuit file FIG36-1. Bring down the Bode plotter enlargement and make sure that the following settings are selected: *Magnitude*, Vertical (Log, F = 50 dB, I = 0 dB), Horizontal (Log, F = 10 kHz, I = 1 Hz). You will plot the frequency response curve for the amplifier gain in dB as the frequency varies between 1 Hz and 10 kHz using the Bode plotter. Notice that the amplifier is identical to the one studied in Experiment 17.

NOTE: If you are performing this experiment in a lab environment, you may not have a Bode plotter available. You will need to plot the frequency response curve by making dB gain measurements at different frequencies between 1 Hz and 10 kHz and plotting the curve on semilog graph paper.

Step 2.   Run the simulation. Notice that the low-frequency response curve has been plotted between the frequencies of 1 Hz and 10 kHz by the Bode plotter. Move the cursor to the flat part of the curve at a frequency of approximately 5 kHz and measure the midfrequency voltage gain in dB. Record your answer.

A_{dB} = _____ dB

Step 3.   Calculate the actual voltage gain from the dB gain measured in Step 2.

Question: How did the voltage gain calculated from the dB gain compare with the voltage gain measured in Experiment 17, Steps 2 and 3?

Step 4.   Move the cursor as close as possible to a point on the curve that is 3 dB down from the dB gain measured in Step 2. Record the frequency (low cutoff frequency, f_{CL}) and the dB gain.

f_{CL} = _____ A_{dB} = _____ dB

Step 5.   Move the cursor to a point on the curve where the frequency is as close as possible to one-tenth the frequency measured in Step 4. Record the new dB gain and the frequency (f).

A_{dB} = _____ dB   f = _____

Question: Approximately how much did the dB gain decrease for a one decade drop in frequency in Steps 4 and 5?   Was this what you expected?

Step 6.   Click *Phase* on the Bode plotter to plot the phase curve. Change the vertical axis initial value (I) to $-180°$ and the final value (F) to $0°$. Run the simulation again. You are looking at the phase difference between the amplifier input and output waveshapes as a function of frequency.

Step 7.   Move the cursor to approximately 10 kHz on the curve plot (midfrequency). Record the phase in degrees. Move the cursor as close as possible on the curve to a frequency that is equal to the value measured in Step 4 (low cutoff frequency). Record the phase in degrees. Also record the difference between the two phase readings.

            Phase = _____ degrees (at mid-frequency)

            Phase = _____ degrees (at low cutoff frequency)

            Phase difference = _____ degrees

Step 8.   Based on the amplifier component values, calculate the cutoff frequency for the input coupling network (f_{in}), the output coupling network (f_o), and the emitter bypass network (f_e). Use the value for the amplifier input resistance (R_{in}) measured in Experiment 17, Step 8. Use the value for the transistor emitter resistance (r_e) measured in Experiment 11, Step 12. Assume that the resistance of the signal source (R_S) is negligible. *Note which network is dominant.*

Questions: Which network was the dominant network?   What was the cutoff frequency of this network?

How did the amplifier low cutoff frequency in Step 4 compare with the cutoff frequency of the dominant network?   **Explain.**

Step 9.   Open circuit file FIG36-2. The circuit is identical to the circuit in Figure 36-1, except that
               the oscilloscope has replaced the Bode plotter. Bring down the oscilloscope enlargement and
               make sure that the following settings are selected: Time base (Scale = 100 μs/Div, Xpos = 0,
               Y/T), Ch A (Scale = 1 mV/Div, Ypos = 0, AC), Ch B (Scale = 200 mV/Div, Ypos = 0, AC),
               Trigger (Pos edge, Level = 0 V, Sing, A). Bring down the function generator enlargement
               and make sure that the following settings are selected: *Sine Wave*, Freq = 5 kHz, Ampl = 2
               mV, Offset = 0. Run the simulation and record the ac peak-to-peak input voltage (V_{in}) and
               the ac peak-to-peak output voltage (V_o).

               V_{in} = _____ V_o = _____

Step 10.   Keep reducing the frequency of the function generator and running the analysis until the ac
               peak-to-peak output voltage (V_o) reaches a value equal to 0.707 times the output voltage
               measured in Step 9. Adjust the oscilloscope as needed. Record the frequency (f_{CL}). This
               should be close to the low cutoff frequency measured in Step 4.

               f_{CL} = _____

Step 11.   Change capacitor C_3 to 20 μF and find the new low cutoff frequency (f_{CL}) following the
               procedure in Step 10. Record your answer.

               f_{CL} = _____

Question: What was the effect of decreasing the emitter bypass capacitance on the amplifier low cutoff
frequency? **Explain.**

Step 12.   Calculate the new cutoff frequency for the emitter bypass network . Note the new dominant
               network.

Question: How did the new dominant network critical frequency in Step 12 compare with the new low
cutoff frequency in Step 11?

37

High-Frequency Amplifier Response

Objectives:

1. Plot the amplifier high-frequency response curve for a common-emitter transistor amplifier.
2. Determine the midfrequency voltage gain from the dB gain for a common-emitter transistor amplifier.
3. Determine the high cutoff frequency for a common-emitter transistor amplifier.
4. Determine the high-frequency voltage gain roll-off in decibels (dB) per decade for a common-emitter transistor amplifier.
5. Determine the phase shift between the input and output waveshapes at the cutoff frequency and at midfrequency for a common-emitter transistor amplifier.
6. Calculate the input bypass network and the output bypass network cutoff frequencies for a common-emitter transistor amplifier.
7. Select the dominant network in order to determine the expected high cutoff frequency, and compare the expected value to the measured value.
8. Demonstrate the effect of changes in signal source resistance on the high cutoff frequency.

Materials:

One 2N3904 bipolar transistor
Capacitors: one 1.0 μF, one 10 μF, one 470 μF
One 20 V dc power supply
One dual-trace oscilloscope
One function generator
Resistors: one 1 Ω, one 100 Ω, one 660 Ω, two 2 kΩ, one 10 kΩ, one 200 kΩ

Theory:

Voltage amplifier **dB voltage gain (A_{dB})** is calculated from the actual voltage gain (A_V) as follows:

$$A_{dB} = 20 \log A_V$$

If A_V is greater than one, the dB voltage gain is positive. If A_V is equal to one, the dB voltage gain is zero. If A_V is less than one, the dB voltage gain is negative and is referred to as attenuation.

The **cutoff frequency (f_C)** on a gain-frequency plot is the frequency where the voltage gain has dropped by **3 dB (0.707)** from the dB gain at midfrequency. At this frequency (f_C), the output power is one-half its midfrequency value. The difference between the high and low cutoff frequencies determines the **bandwidth** of an amplifier.

At low frequencies, **coupling and bypass capacitors** affect the voltage gain of an amplifier because the capacitive reactances are significant at these frequencies. At the midrange frequencies, the effects of capacitances are minimal and can be neglected. At high frequencies, the coupling and bypass capacitors have no effect on ac performance because their high capacitance values produce low reactances. However, **internal device capacitances** do affect performance at high frequencies, reducing the gain and introducing phase shift as the signal frequency increases. At low frequencies, the internal device capacitances have no effect on amplifier performance because of their very low capacitance values.

In this experiment, you will examine how amplifier voltage gain and phase shift are affected when the signal frequency approaches the amplifier **high cutoff frequency**. You will use the Bode plotter to study the **high-frequency amplifier response** of the **small-signal common-emitter amplifier** in Experiment 17. The amplifier circuit in Figure 17-1, Experiment 17, has been repeated in Figures 37-1 and 37-2.

Because capacitive reactance decreases with increasing frequency, internal transistor capacitances can no longer be considered open circuits at high frequencies. In Figure 37-1, the **transistor base-emitter junction capacitance (C_{be})**, the **base-collector junction capacitance (C_{bc})**, and resistors R_1, R_2, R_S, and r_e form an **input low-pass filter** with a **cutoff frequency (f_{in})** that can be calculated from

$$f_{in} = \frac{1}{2\pi R C_{in}}$$

where $R = R_S \| R_1 \| R_2 \| \beta r_e \cong R_S$

$C_{in} = C_{be} + C_{bc} A_V$

In Figure 37-1, **transistor base-collector junction capacitance (C_{bc})** and output resistances R_C and R_L form an **output low-pass filter** with a **cutoff frequency (f_o)** that can be calculated from

$$f_o = \frac{1}{2\pi R_c C_o}$$

where $R_c = R_C R_L /(R_C + R_L)$

$C_o = C_{bc} (A_V + 1)/A_V \cong C_{bc}$

The voltage gain will drop **20 dB per decade** of frequency increase above the cutoff frequency for each low-pass filter network. The **phase** difference between the input and the output will shift by **45 degrees** at the cutoff frequency from what it was at midfrequency for each single-pole network.

The high cutoff frequency for the amplifier is determined by establishing which of the two low-pass filters has the lowest cutoff frequency. This network is called the **dominant network** and its cutoff frequency determines the high cutoff frequency for the amplifier (voltage gain 3 dB down from midfrequency gain).

Figure 37-1 Common-Emitter Amplifier High-Frequency Response

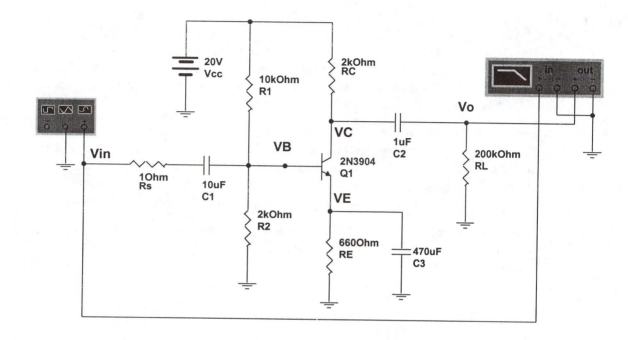

Figure 37-2 Common-Emitter Amplifier High-Frequency Response

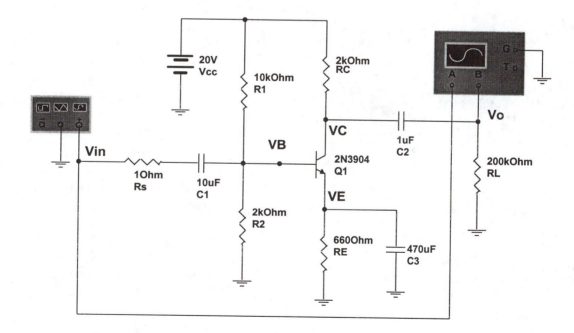

Procedure:

Step 1.   Open circuit file FIG37-1. Bring down the Bode plotter enlargement and make sure that the following settings are selected: *Magnitude*, Vertical (Log, F = 50 dB, I = 0 dB), Horizontal (Log, F = 500 MHz, I = 10 kHz). You will plot the frequency response curve for the amplifier gain in dB as the frequency varies between 10 kHz and 500 MHz, using the Bode plotter. Notice that the amplifier is identical to the one studied in Experiment 17, except for the 1 Ω resistor (R_S) in series with the signal source.

> *NOTE:* If you are performing this experiment in a lab environment, you may not have a Bode plotter available. You will need to plot the frequency response curve by making dB gain measurements at different frequencies between 10 kHz and 500 MHz and plotting the curve on semilog graph paper.

Step 2.   Run the simulation. Notice that the high-frequency response curve has been plotted between the frequencies of 10 kHz and 500 MHz by the Bode plotter. Move the cursor to the flat part of the curve at a frequency of approximately 20 kHz and measure the midfrequency voltage gain in dB. Record your answer.

$$A_{dB} = \underline{\hspace{2cm}} dB$$

Step 3.   Calculate the actual voltage gain from the dB gain measured in Step 2.

Question: How did the voltage gain calculated from the dB gain compare with the voltage gain measured in Experiment 17, Steps 2 and 3?

Step 4.   Move the cursor as close as possible to a point on the curve that is 3 dB down from the dB gain measured in Step 2. Record the frequency (high cutoff frequency, f_{CH}) and the dB gain.

$$f_{CH} = \underline{\hspace{2cm}} \qquad\qquad A_{dB} = \underline{\hspace{2cm}} dB$$

Step 5.   Move the cursor to a point on the curve where the frequency is as close as possible to ten times f_{CH}. Record the dB gain (A_{dB}) and frequency (f).

$$A_{dB} = \underline{\hspace{2cm}} dB \qquad\qquad f = \underline{\hspace{2cm}}$$

Question: Approximately how much did the dB gain decrease for a one decade increase in frequency? Was this what you expected?

Step 6.   Click *Phase* on the Bode plotter to plot the phase curve. Change the vertical axis initial value (I) to 0° and the final value (F) to 180°. Run the simulation again. You are looking at the phase difference between the amplifier input and output waveshapes as a function of frequency.

Step 7.   Move the cursor to approximately 40 kHz on the flat part of the curve plot (midfrequency). Record the phase in degrees. Move the cursor as close as possible on the curve to a frequency that is equal to the value measured in Step 4 (high cutoff frequency). Record the phase in degrees.

Phase = _____ degrees (at mid-frequency)

Phase = _____ degrees (at high cutoff frequency)

Question: What was the difference between the midfrequency phase and the high cutoff frequency phase? Was it close to what you expected for a single-pole network?

Step 8.   Based on the amplifier component values, calculate the critical frequency for the input low-pass network (f_{in}) and the output low-pass network (f_o). Use the values of the base-emitter junction capacitance and the base-collector junction capacitance in the 2N3904 transistor model library to estimate the values of C_{be} (C_{je}) and C_{bc} (C_{jc}), respectively. These values can be obtained by double-clicking the 2N3904 transistor and clicking *Edit Model*. Use the value for the transistor emitter resistance (r_e) measured in Experiment 11, Step 12 and the value of β measured in Experiment 11, Step 10. Consider the resistance of the signal source (R_S) to be equal to the 1 Ω resistor in series with the source. **Note which network is dominant.**

Questions: Which network was the dominant network?   What was the cutoff frequency of this network?

How did the amplifier high cutoff frequency (f_{CH}) in Step 4 compare with the cutoff frequency of the dominant network? **Explain.**

Step 9.   Open circuit file FIG37-2. The circuit is identical to the circuit in Figure 37-1, except that the oscilloscope has replaced the Bode plotter. Bring down the oscilloscope enlargement and make sure that the following settings are selected: Time base (Scale = 20 μs/Div, Xpos = 0, Y/T), Ch A (Scale = 1 mV/Div, Ypos = 0, AC), Ch B (Scale = 200 mV/Div, Ypos = 0, AC), Trigger (Pos edge, Level = 0 V, Sing, A). Bring down the function generator enlargement and make sure that the following settings are selected: *Sine Wave*, Freq = 20 kHz, Ampl = 2 mV, Offset = 0. Run the simulation and record the ac peak-to-peak input voltage (V_{in}) and the ac peak-to-peak output voltage (V_o).

V_{in} = _____ V_o = _____

Step 10.   Keep increasing the frequency of the function generator and running the simulation until the ac peak-to-peak output voltage (V_o) reaches a value equal to 0.707 times the output voltage at midfrequency (the output voltage recorded in Step 9). Adjust the oscilloscope as needed. Record this frequency (f_{CH}). This should be close to the high cutoff frequency measured in Step 4.

f_{CH} = _____

Step 11.   Change the source resistance (R_S) to 100 Ω and set the function generator frequency to 20 kHz (midfrequency). Run the simulation and record the ac peak-to-peak input voltage (V_{in}) and the ac peak-to-peak output voltage (V_o). Adjust the oscilloscope as needed.

V_{in} = _____ V_o = _____

Step 12.   Based on the values measured in Step 11, calculate the new midfrequency voltage gain (A_V).

Step 13.   Following the procedure in Step 10, determine the new high cutoff frequency (f_{CH}). Record your answer. Adjust the oscilloscope as needed.

f_{CH} = _____

Question: What was the effect of increasing the source resistance (R_S) on the amplifier high cutoff frequency? **Explain.**

Step 14.   Calculate the new critical frequency for the input low-pass filter. **Note the new dominant network.** (Don't forget to use the new value for R_S and the new voltage gain measured in Steps 11 and 12).

Question: How did the new dominant network cutoff frequency in Step 14 compare with the new cutoff frequency measured in Step 13?

38

Op-Amp Frequency Response

Objectives:

1. Plot the frequency response curve for an open-loop and a closed-loop operational amplifier.
2. Determine the midfrequency voltage gain from the dB gain for an open-loop and a closed-loop operational amplifier.
3. Determine the high cutoff frequency (bandwidth) for an open-loop and a closed-loop operational amplifier.
4. Determine the high-frequency voltage gain roll-off in dB per decade for an open-loop operational amplifier.
5. Determine the unity-gain bandwidth for an operational amplifier.
6. Calculate the amplifier bandwidth using the op-amp unity-gain bandwidth and compare the measured value with the calculated value.
7. Determine the phase shift between the input and output waveshapes at the cutoff frequency and at midfrequency for an open-loop operational amplifier.
8. Determine the effect of negative feedback on bandwidth for a closed-loop operational amplifier.
9. Determine the high cutoff frequency (bandwidth) using a pulse input for a closed-loop operational amplifier.

Materials:

Two 15 V dc voltage supplies
One LM741 op-amp
One function generator
One dual-trace oscilloscope
Resistors: one 1 kΩ, one 10 kΩ, two 100 kΩ

Theory:

Voltage amplifier **dB voltage gain** is calculated from the actual voltage gain (A_V) as follows:

$$A_{dB} = 20 \log A_V$$

If A_V is greater than one, the dB voltage gain is positive. If A_V is equal to one, the dB voltage gain is zero. If A_V is less than one, the dB voltage gain is negative and is referred to as **attenuation**.

The **cutoff frequency (f_C)** on a **gain-frequency plot** is the frequency where the voltage gain has dropped by **3 dB (0.707)** from the dB gain at midfrequency. At this frequency (f_C), the output power is one-half its midfrequency value. The difference between the high and low cutoff frequencies determines the **bandwidth** of an amplifier. Because op-amps are **dc amplifiers**, there is no low cutoff frequency. Therefore, the midrange gain extends down to zero frequency, making the bandwidth equal to the **high cutoff frequency (f_{CH})**. As the frequency is increased beyond the high cutoff frequency, the voltage gain will continue to drop until it reaches unity (0 dB). The frequency at which the voltage gain is one (0 dB) is called the **unity-gain frequency (f_{unity})**. The product of the amplifier midfrequency voltage gain (A_V) and the bandwidth (f_{CH}) is called the **gain-bandwidth product** and will always be equal to the unity-gain frequency (f_{unity}) when the gain roll-off is **–20 dB per decade**. Therefore,

$$A_V \, f_{CH} \; = \; f_{unity}$$

or

$$f_{CH} = \frac{f_{unity}}{A_V}$$

This means that the bandwidth of an op-amp amplifier with a voltage gain A_V will have a bandwidth (f_{CH}) that is inversely proportional to the voltage gain. The bandwidth (f_{CH}) for an open-loop op-amp is designed to be very narrow because a wider bandwidth with such a high voltage gain would cause the amplifier to **oscillate**. As the voltage gain is reduced with negative feedback, the bandwidth will become wider. The **phase difference** between the input and the output will change as the frequency is increased. The phase at the cutoff frequency will be **45 degrees** from what it was at midfrequency when the gain roll-off is –20 dB per decade.

The high cutoff frequency (bandwidth) and the unity-gain frequency (f_{unity}) for an open-loop op-amp will be measured using the circuit in Figure 38-1. The high cutoff frequency (bandwidth) for a closed-loop noninverting op-amp voltage amplifier will be measured using the circuit in Figure 38-2.

As previously demonstrated, the voltage gain (A_V) of the noninverting op-amp voltage amplifier in Figure 38-2 is calculated from

$$A_V = 1 + \frac{R_1}{R_2}$$

If a square wave is applied to the input of a noninverting op-amp voltage amplifier, shown in Figure 38-3, the risetime (T_r) of the amplifier output square wave (from the 10% point to the 90% point) can be used to determine the amplifier bandwidth (f_{CH}). The relationship between the output risetime and the amplifier bandwidth is

$$f_{CH} = \frac{0.35}{T_r}$$

Figure 38-1 Open-Loop Op-Amp Frequency Response

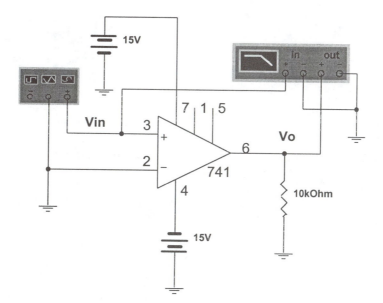

Figure 38-2 Noninverting Voltage Amplifier Frequency Response

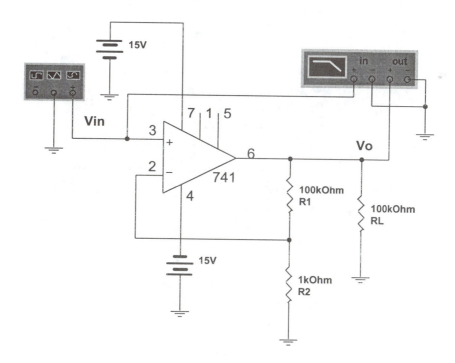

Figure 38-3 Noninverting Voltage Amplifier Pulse Response

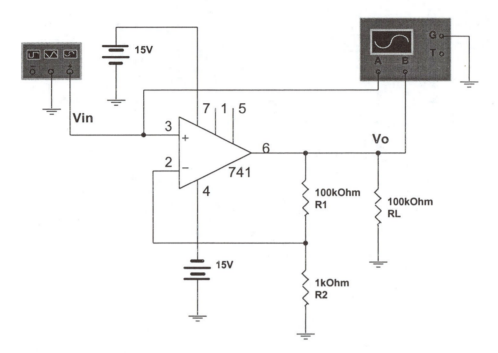

Procedure:

Step 1.   Open circuit file FIG38-1. Bring down the Bode plotter enlargement and make sure that the following settings are selected: *Magnitude*, Vertical (Log, F = 110 dB, I = 0 dB), Horizontal (Log, F = 2 MHz, I = 1 Hz). You will plot the frequency response curve for the op-amp open-loop voltage gain in dB as the frequency varies between 1.0 Hz and 2 MHz using the Bode plotter.

NOTE: If you are performing this experiment in a lab environment, you may not have a Bode plotter available. You will need to plot the frequency response curve by making dB gain measurements at different frequencies between 1.0 Hz and 2 MHz and plotting the curve on semilog graph paper.

Step 2.   Run the simulation. Notice that the frequency response curve has been plotted between the frequencies of 1.0 Hz and 2 MHz by the Bode plotter. Move the cursor to the flat part of the curve at a frequency of approximately 1 Hz and measure the midfrequency open-loop voltage gain in dB. Record your answer.

A_{dB} = _____ dB (open-loop)

Step 3.   Calculate the open-loop voltage gain (A_V) from the dB gain measured in Step 2.

Question: How did the open-loop voltage gain calculated from the dB gain compare with the expected open-loop voltage gain for a 741 op-amp?

Step 4.   Move the cursor as close as possible to a point on the curve that is 3 dB down from the dB gain measured in Step 2. Record the high cutoff frequency f_{CH} (bandwidth) and the dB gain.

f_{CH} = _____ A_{dB} = _____ dB

Question: Why was the bandwidth so low for the open-loop op-amp?

Step 5.   Move the cursor to a point on the curve where the frequency is as close as possible to ten times f_{CH}. Record the new dB gain (A_{dB}) and frequency (f_1).

A_{dB} = _____ dB f_1 = _____

Question: Approximately how much did the dB gain decrease for a one decade increase in frequency for the open-loop op-amp?   Was this what you expected?

Step 6.   Move the cursor to the point on the curve where it crosses the 0 dB line. Record the frequency. This frequency is the value of the unity-gain bandwidth (f_{unity}) for the op-amp. (Remember that 0 dB represents a voltage gain of one.)

f_{unity} = _____

Question: How did the op-amp unity-gain bandwidth compare with the value expected for a 741 op-amp?

Step 7.   Calculate the expected bandwidth (f_{CH}) for the open-loop op-amp based on the unity-gain bandwidth (f_{unity}) measured in Step 6 and the gain calculated in Step 3.

Question: How did the calculated bandwidth for the open-loop op-amp compare with the measured bandwidth?

Step 8.   Click *Phase* on the Bode plotter to plot the phase curve. Change the vertical axis initial value (I) to –90°, the final value (F) to 0°, and the horizontal axis I to 0.1 Hz (100 mHz). Run the simulation again. You are looking at the phase difference between the amplifier input and output waveshapes as a function of frequency.

Step 9.   Move the cursor to the top of the curve plot (midfrequency). Record the phase in degrees. Move the cursor as close as possible on the curve to a frequency that is equal to the value measured in Step 4 (high cutoff frequency). Record the phase in degrees.

    Phase = _____ degrees (at midfrequency)

    Phase = _____ degrees (at high cutoff frequency)

Question: For the open-loop op-amp, what was the difference between the midfrequency phase and the high cutoff frequency phase? Was it close to what you expected?

Step 10.   Open circuit file FIG38-2. Bring down the Bode plotter enlargement and make sure that the following settings are selected: *Magnitude*, Vertical (Log, F = 50 dB, I = 0 dB), Horizontal (Log, F = 2 MHz, I = 1 Hz). You will plot the frequency response curve for the op-amp closed-loop voltage gain in dB as the frequency varies between 1.0 Hz and 2 MHz using the Bode plotter.

Step 11.   Run the simulation. Notice that the frequency response curve has been generated between the frequencies of 1.0 Hz and 2 MHz by the Bode plotter. Move the cursor to the flat part of the curve at a frequency of approximately 1 Hz. Measure the midfrequency closed-loop voltage gain in dB. Record your answer.

    A_{dB} = _____ dB (closed-loop)

Step 12.   Calculate the amplifier closed-loop voltage gain (A_V) from the dB voltage gain measured in Step 11.

Step 13.   Calculate the amplifier closed-loop voltage gain (A_V) based on the value of the circuit components.

Question: How did your calculated amplifier closed-loop voltage gain based on the amplifier component values compare with the value determined from the dB voltage gain in Step 12?

Step 14.   Move the cursor as close as possible to a point on the curve that is 3 dB down from the dB voltage gain measured in Step 11. Record the high cutoff frequency, f_{CH} (bandwidth) and the dB gain.

   f_{CH} = _____ A_{dB} = _____ dB

Step 15.   Calculate the expected bandwidth (f_{CH}) for the closed-loop op-amp amplifier based on the op-amp unity-gain bandwidth (f_{unity}) measured in Step 6 and the amplifier voltage gain calculated in Step 13.

Question: How did your calculated bandwidth for the closed-loop amplifier compare with the measured bandwidth?

Step 16.   Change resistor R_1 to 10 kΩ and repeat Step 11.

   A_{dB} = _____ dB (closed loop)

Step 17.   Calculate the amplifier closed-loop voltage gain (A_V) from the dB voltage gain measured in Step 16.

Step 18.   Calculate the amplifier closed-loop voltage gain (A_V) based on the value of the circuit components.

Question: How did your calculated amplifier closed-loop voltage gain based on the amplifier component values compare with the value determined from the dB gain in Step 17?

Step 19.   Move the cursor as close as possible to a point on the curve that is 3 dB down from the dB gain measured in Step 16. Record the high cutoff frequency f_{CH} (bandwidth) and the dB gain.

f_{CH} = _____ A_{dB} = _____ dB

Question: What effect did reducing the closed-loop voltage gain have on the amplifier bandwidth? **Explain.**

Step 20.   Calculate the expected bandwidth (f_{CH}) for the closed-loop op-amp amplifier based on the op-amp unity-gain bandwidth (f_{unity}) measured in Step 6 and the voltage gain calculated in Step 18.

Question: How did your calculated value for f_{CH} compare with the value measured in Step 19?

Step 21.　Open circuit file FIG38-3. The circuit is identical to the circuit in Figure 38-2, except that the oscilloscope has replaced the Bode plotter. Bring down the oscilloscope enlargement and make sure that the following settings are selected: Time base (Scale = 10 μs/Div, Xpos = 0, Y/T), Ch A (Scale = 5 mV/Div, Ypos = 0, DC), Ch B (Scale = 500 mV/Div, Ypos = 0, DC), Trigger (Pos edge, Level = 0 V, Sing, A). Run the simulation. Measure the output pulse (blue curve) rise-time (T_r) from the 10% point to the 90% point. Record your answer.

T_r = _____

Step 22.　Calculate the amplifier bandwidth (f_{CH}) from the risetime (T_r) measured in Step 21.

Question: How did the bandwidth calculated using the pulse input compare with the bandwidth measured in Step 14?

PART

VI

Active Filters

The following four experiments involve plotting frequency response curves and measuring the cutoff frequencies of active filters. First, you will plot the frequency response curve and determine the cutoff frequency for a first-order and a second-order low-pass filter. Next, you will plot the frequency response curve and determine the cutoff frequency for a first-order and a second-order high-pass filter. Finally, you will plot the frequency response curve and determine the bandwidth of a band-pass filter and a band-stop (notch) filter.

The circuits for the experiments in Part VI can be found on the enclosed disk in the FILTERS subdirectory.

EXPERIMENT

39

Name_____

Date_____

Low-Pass Active Filters

Objectives:

1. Plot the frequency response curve for a first-order (one-pole) low-pass active filter.
2. Determine the cutoff frequency for a first-order low-pass active filter.
3. Determine the roll-off in decibels (dB) per decade for a first-order low-pass active filter.
4. Plot the phase response curve as a function of frequency for a first-order low-pass active filter.
5. Plot the frequency response curve for a second-order (two-pole) low-pass active filter.
6. Determine the cutoff frequency for a second-order low-pass active filter.
7. Determine the roll-off in decibels (dB) per decade for a second-order low-pass active filter.
8. Plot the phase response curve as a function of frequency for a second-order low-pass active filter.

Materials:

Two 15 V dc voltage supplies
One LM741 op-amp
Capacitors: two 0.001 μF, one 1 pF
One function generator
One dual-trace oscilloscope
Resistors: one 1 kΩ, one 5.86 kΩ, two 10 kΩ, two 33 kΩ

Theory:

Filters are circuits that pass selected frequencies while rejecting other frequencies. **Active filters** use **active devices** such as op-amps combined with **passive elements** such as resistors, capacitors, and inductors. The passive elements provide **frequency selectivity** and the active devices provide **voltage gain**, **high input impedance**, and **low output impedance**. **Active filters** have several advantages over **passive filters** (R, C, and L elements only). The voltage gain reduces attenuation of the signal by the filter, the high input impedance prevents excessive loading of the source, and the low output impedance prevents the filter from being affected by the load. Active filters are also easy to adjust over a wide frequency range without altering the desired response. The weakness of active filters is that the cutoff frequency cannot exceed the **unity-gain bandwidth (f_{unity})** of the op-amp. Therefore, active filters are mostly used in **low-frequency applications**.

The **ideal filter** has an instantaneous roll-off at the **cutoff frequency (f_C)**, with full signal level on one side of the cutoff frequency and no signal level on the other side of the cutoff frequency. Although the ideal is not achievable, actual filters roll off at **–20 dB/decade or higher**, depending on the type of filter.

321

The –20 dB/decade roll-off is obtained with a **one-pole filter** (one RC circuit). A **two-pole filter** has two RC circuits tuned to the same cutoff frequency and rolls off at **–40 dB/decade**. Each additional pole (RC circuit) will cause the filter to roll off an additional –20 dB/decade. In a one-pole filter, the **phase** between the input and the output will change by **90 degrees** over the frequency range and be **45 degrees** at the cutoff frequency. In a two-pole filter, the phase will change by **180 degrees** over the frequency range and be **90 degrees** at the cutoff frequency.

The three basic types of response characteristics that can be realized with most active filters are Butterworth, Chebyshev, and Bessel, depending on the selection of certain filter component values. The **Butterworth filter** provides a flat amplitude response in the pass band and a roll-off of –20 dB/decade/pole with a nonlinear phase response. The **Chebyshev filter** provides a ripple amplitude response in the pass band and a roll-off better than –20 dB/decade/pole with a less linear phase response than the Butterworth filter. The **Bessel filter** provides a flat amplitude response in the pass band and a roll-off of less than –20 dB/decade/pole with a linear phase response. Because of its maximally flat response in the pass band, the Butterworth filter is the most widely used.

The four basic types of active filters are low-pass, high-pass, band-pass, and band-reject. A **low-pass filter** passes frequencies below the cutoff frequency and attenuates frequencies above the cutoff frequency. The **cutoff frequency** on the gain-frequency plot is the frequency where the voltage gain has dropped by **3 dB (0.707)** from the dB gain at the low frequencies. In this experiment, you will study the low-pass active filters in Figures 39-1 and 39-2. The expected **cutoff frequency (f_c)** for the **one-pole low-pass filter** in Figure 39-1 and the **two-pole Sallen-Key low-pass Butterworth filter** in Figure 39-2 can be calculated from

$$f_c = \frac{1}{2\pi RC}$$

The two-pole Sallen-Key low-pass Butterworth filter in Figure 39-2 requires a **voltage gain of 1.586** to produce the Butterworth response. Therefore,

$$A_V = 1 + \frac{R_1}{R_2} = 1.586$$

and

$$\frac{R_1}{R_2} = 0.586$$

The filter network **dB voltage gain** is calculated from the actual voltage gain (A_V) as follows:

$$A_{dB} = 20 \log A_V$$

Figure 39-1 First-Order (1-pole) Low-Pass Active Filter

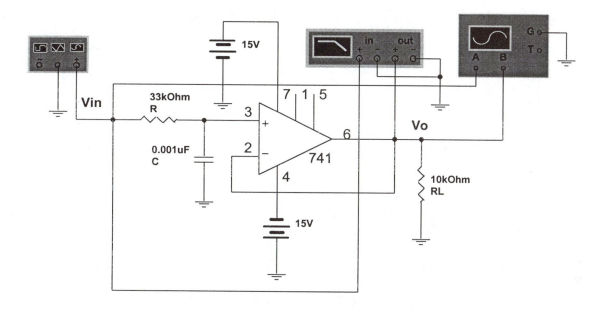

Figure 39-2 Second-Order (2-pole) Sallen-Key Low-Pass Butterworth Filter

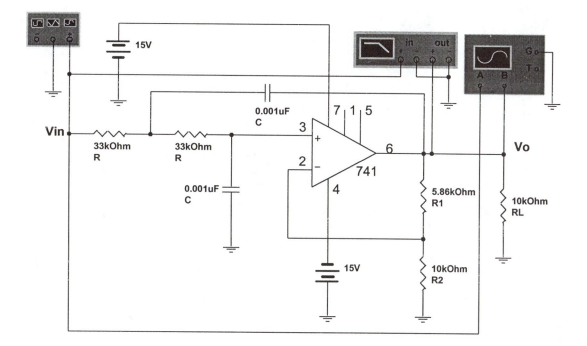

Procedure:

Step 1.   Open circuit file FIG39-1. Bring down the Bode plotter enlargement and make sure that the following settings are selected: *Magnitude*, Vertical (Log, F = 0 dB, I = –40 dB), Horizontal (Log, F = 1 MHz, I = 100 Hz). You will plot the frequency response curve for a first-order (one-pole) low-pass active filter as the frequency varies between 100 Hz and 1 MHz using the Bode plotter.

NOTE: If you are performing this experiment in a lab environment, you may not have a Bode plotter available. You will need to plot the frequency response curve by making dB voltage gain measurements at different frequencies between 100 Hz and 1 MHz and plotting the curve on semilog graph paper.

Step 2.   Run the simulation. Notice that the frequency response curve has been plotted between the frequencies of 100 Hz and 1 MHz by the Bode plotter. Draw the curve plot in the space provided. Next, move the cursor to the flat part of the curve at a frequency of approximately 100 Hz and measure the voltage gain in dB. *Record the dB voltage gain on the curve plot.*

One-Pole Low-Pass Filter

A_{dB}

f

Question: Is the frequency response curve that of a low-pass filter? **Explain.**

Step 3.   Calculate the actual voltage gain (A_V) from the dB voltage gain measured in Step 2.

Question: Was the voltage gain (A_V) on the flat part of the frequency response curve what you expected for the circuit in Figure 39-1? **Explain why.**

Step 4.   Move the cursor as close as possible to a point on the curve that is 3 dB down from the dB gain at the low frequencies measured in Step 2. Record the frequency (cutoff frequency, f_C) on the curve plot drawn in Step 2.

Step 5.   Calculate the expected cutoff frequency (f_C) based on the circuit component values.

Question: How did the calculated value for the cutoff frequency compare with the measured value recorded on the curve plot for the one-pole low-pass filter?

Step 6.   Move the cursor to a point on the curve where the frequency is as close as possible to ten times f_C. Record the new dB voltage gain and frequency (f_2) on the curve plot drawn in Step 2.

Questions: Approximately how much did the dB voltage gain decrease for a one decade increase in frequency for the one-pole low-pass filter?   Was this what you expected for a one-pole filter?

Step 7.   Click *Phase* on the Bode plotter to plot the phase curve. Change the vertical axis initial
           value (I) to –90° and the final value (F) to 0°. Run the simulation again. You are looking at
           the phase difference between the filter input and output waveshapes as a function of
           frequency. Draw the curve plot in the space provided.

One-Pole Low-Pass Filter

ϕ

f

Step 8.   Move the cursor to the top of the curve plot (low frequency). Record the phase in degrees on
           the curve plot. Move the cursor as close as possible on the curve to a frequency that is equal
           to the value measured in Step 4 (cutoff frequency, f_C). Record the frequency and phase (ϕ)
           on the curve plot.

Question: Was the phase shift between input and output at the cutoff frequency what you expected for a
one-pole low-pass filter?

Step 9.   Bring down the oscilloscope enlargement and make sure that the following settings are
           selected: Time base (Scale = 500 µs/Div, Xpos = 0, Y/T), Ch A (Scale = 1 V/Div, Ypos = 0,
           DC), Ch B (Scale = 500 mV/Div, Ypos = 0, DC), Trigger (Pos edge, Level = 0 V, Auto).
           Bring down the function generator enlargement and make sure that the following settings
           are selected: *Sine Wave*, Freq = 1 kHz, Ampl = 1 V, Offset = 0. Run the simulation and keep
           increasing the function generator frequency. Notice what happens to the output sine wave
           magnitude and phase as you approach the cutoff frequency (f_C). Adjust the oscilloscope as
           needed. Record the peak-to-peak output voltage magnitude (V_o) and the phase (ϕ) between
           the input and output at f = 1 kHz and at f = f_C.

V_o = _____ ϕ = _____   (at f = 1 kHz)

V_o = _____ ϕ = _____   (at f = f_C)

Questions: Did the output magnitude do what you expected for a one-pole low-pass filter when you approached the cutoff frequency? Did it drop 3 dB (0.707) at the cutoff frequency?

Did the phase between the input and output change as you expected for a one-pole filter as you approached the cutoff frequency?

Step 10.   Return the oscilloscope and function generator to normal size to expose the Bode plotter. Click *Magnitude* on the plotter. Change R to 1 kΩ and C to 1 pF. Adjust the horizontal final frequency (F) on the Bode plotter to 20 MHz. Run the simulation and measure the cutoff frequency (f_C) following the procedure in Step 4. Record your answer.

f_C = _____

Step 11.   Based on the new values for resistor R and capacitor C, calculate the new cutoff frequency (f_C).

Question: Explain why there was such a large difference between the calculated and the measured values for the cutoff frequency in Steps 10 and 11 when R = 1 kΩ and C = 1 pF. *Hint:* What was the value of the unity-gain bandwidth, f_{unity}, for the 741 op-amp?

Step 12.   Open circuit file FIG39-2. Bring down the Bode plotter enlargement and make sure that the following settings are selected: *Magnitude*, Vertical (Log, F = 10 dB, I = –40 dB), Horizontal (Log, F = 1 MHz, I = 100 Hz). You will plot the frequency response curve for a second-order (two-pole) low-pass active filter as the frequency varies between 100 Hz and 1 MHz using the Bode plotter.

NOTE: If you are performing this experiment in a lab environment, you may not have a Bode plotter available. You will need to plot the frequency response curve by making dB voltage gain measurements at different frequencies between 100 Hz and 1 MHz and plotting the curve on semilog graph paper.

Step 13.   Run the simulation. Notice that the frequency response curve has been plotted between the frequencies of 100 Hz and 1 MHz by the Bode plotter. Draw the curve plot in the space provided. Next, move the cursor to the flat part of the curve at a frequency of approximately 100 Hz and measure the voltage gain in dB. *Record the dB voltage gain on the curve plot.*

Two-Pole Low-Pass Filter

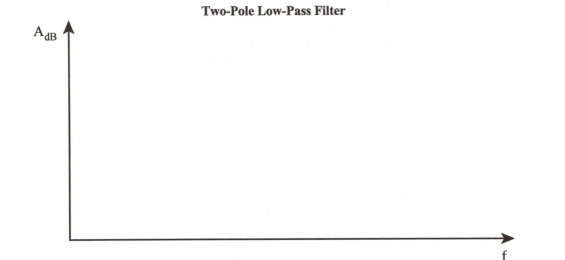

Question: Is the frequency response curve that of a low-pass filter? **Explain.**

Step 14.   Calculate the actual voltage gain (A_V) from the dB voltage gain measured in Step 13.

Question: Was the voltage gain (A_V) on the flat part of the frequency response curve what you expected for the Butterworth filter in Figure 39-2? **Explain.**

Step 15.   Move the cursor as close as possible to a point on the curve that is 3 dB down from the dB voltage gain at the low frequencies measured in Step 13. Record the frequency (cutoff frequency, f_C) on the curve plot drawn in Step 13.

Step 16.   Calculate the expected cutoff frequency (f_C) based on the circuit component values.

Question: How did the calculated value for the cutoff frequency compare with the measured value recorded on the curve plot for the two-pole low-pass active filter?

Step 17.   Move the cursor to a point on the curve where the frequency is as close as possible to ten times f_C. Record the new dB voltage gain and frequency (f_2) on the curve plot drawn in Step 13.

Question: Approximately how much did the dB voltage gain decrease for a one decade increase in frequency for the two-pole low-pass filter?   Was this what you expected for a two-pole filter?

Step 18.   Click *Phase* on the Bode plotter to plot the phase curve. Change the vertical axis initial value (I) to $-180°$ and the final value (F) to $0°$. Run the simulation again. You are looking at the phase difference between the filter input and output waveshapes as a function of frequency. Draw the curve plot in the space provided.

Two-Pole Low-Pass Filter

Step 19.   Move the cursor to the top of the curve plot (low frequency). Record the phase in degrees on the curve plot. Move the cursor as close as possible on the curve to a frequency that is equal to the value measured in Step 15 (cutoff frequency, f_C). Record the frequency and phase (ϕ) on the curve plot drawn in Step 18.

Question: Was the phase shift between input and output at the cutoff frequency what you expected for a two-pole low-pass filter?

Step 20.   Bring down the oscilloscope and function generator and make sure that the settings are the same as in Step 9. Keep running the simulation and increasing the function generator frequency. Notice what happens to the output sine wave magnitude and phase as you approach the cutoff frequency (f_C). Adjust the oscilloscope as needed. Record the peak-to-peak output voltage magnitude (V_o) and the phase (ϕ) between the input and output at $f = 1$ kHz and at $f = f_C$.

$V_o =$ _____ $\phi =$ _____ (at f = 1 kHz)

$V_o =$ _____ $\phi =$ _____ (at f = f_C)

Questions: Did the output magnitude do what you expected for a two-pole low-pass filter when you approached the cutoff frequency in Step 20? Did it drop 3 dB (0.707) at the cutoff frequency?

Did the phase between the input and output change as you expected for a two-pole filter as you approached the cutoff frequency in Step 20?

Step 21.   Return the oscilloscope and function generator to normal size to expose the Bode plotter. Click *Magnitude* on the plotter and change the vertical axis final setting (F) to 50 dB. Change R_1 to 20 kΩ. Run the simulation and draw the curve plot in the space provided.

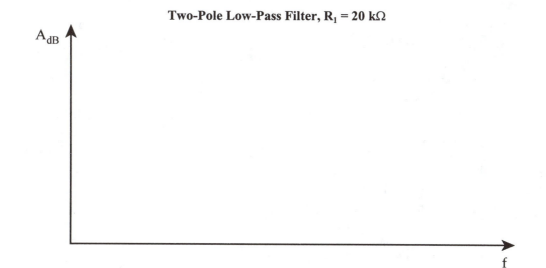

Two-Pole Low-Pass Filter, R_1 = 20 kΩ

Question: Is the curve plot that of a Butterworth filter? What caused the curve plot to change?

Troubleshooting Problems

1.   Open circuit file FIG39-3. Run the simulation. Locate the defective component and state the defect (short or open). You can use any instrument available and make any measurement desired.

           Defective component: _____   Defect: _____

2.   Open circuit file FIG39-4. Run the simulation. Locate the defective component and state the defect (short or open). You can use any instrument available and make any measurement desired.

           Defective component: _____   Defect: _____

3.   Open circuit file FIG39-5. Run the simulation. Locate the defective component and state the defect (short or open). You can use any instrument available and make any measurement desired.

           Defective component: _____   Defect: _____

EXPERIMENT

40

High-Pass Active Filters

Objectives:

1. Plot the frequency response curve for a first-order (one-pole) high-pass active filter.
2. Determine the cutoff frequency for a first-order high-pass active filter.
3. Determine the roll-off in decibels (dB) per decade for a first-order high-pass active filter.
4. Plot the phase response curve as a function of frequency for a first-order high-pass active filter.
5. Plot the frequency response curve for a second-order (two-pole) high-pass active filter.
6. Determine the cutoff frequency for a second-order high-pass active filter.
7. Determine the roll-off in decibels (dB) per decade for a second-order high-pass active filter.
8. Plot the phase response curve as a function of frequency for a second-order high-pass active filter.

Materials:

Two 15 V dc voltage supplies
One LM741 op-amp
One function generator
One dual-trace oscilloscope
Capacitors: two 0.005 μF
Resistors: one 5.86 kΩ, two 10 kΩ, two 30 kΩ

Theory:

Filters are circuits that pass selected frequencies while rejecting other frequencies. **Active filters** use **active devices** such as op-amps combined with **passive elements** such as resistors, capacitors, and inductors. The passive elements provide **frequency selectivity** and the active devices provide **voltage gain**, **high input impedance**, and **low output impedance**. **Active filters** have several advantages over **passive filters** (R, C, and L elements only). The voltage gain reduces attenuation of the signal by the filter, the high input impedance prevents excessive loading of the source, and the low output impedance prevents the filter from being affected by the load. Active filters are also easy to adjust over a wide frequency range without altering the desired response.

An **ideal filter** has an instantaneous roll-off at the **cutoff frequency (f_C)**, with full signal level on one side of the cutoff frequency and no signal level on the other side of the cutoff frequency. Although the ideal is not achievable, actual filters roll off at **–20 dB/decade** or higher, depending on the type of filter. The –20 dB/decade roll-off is obtained with a **one-pole filter** (one RC circuit). A **two-pole filter** has two RC circuits tuned to the same cutoff frequency and rolls off at **–40 dB/decade**. Each additional pole (RC

333

circuit) will cause the filter to roll off an additional –20 dB/decade. In a one-pole filter, the **phase** between the input and the output will change by **90 degrees** over the frequency range and be **45 degrees** at the cutoff frequency. In a two-pole filter, the phase will change by **180 degrees** over the frequency range and be **90 degrees** at the cutoff frequency.

The three basic types of response characteristics that can be realized with most active filters are Butterworth, Chebyshev, and Bessel, depending on the selection of certain filter component values. The **Butterworth filter** provides a flat amplitude response in the pass band and a roll-off of –20 dB/decade/pole with a nonlinear phase response. The **Chebyshev filter** provides a ripple amplitude response in the pass band and a roll-off better than –20 dB/decade/pole with a less linear phase response than the Butterworth filter. The **Bessel filter** provides a flat amplitude response in the pass band and a roll-off of less than –20 dB/decade/pole with a linear phase response. Because of its maximally flat response in the pass band, the Butterworth filter is the most widely used.

The four basic types of active filters are low-pass, high-pass, band-pass, and band-reject. A **high-pass filter** passes frequencies above the cutoff frequency and attenuates frequencies below the cutoff frequency. The **cutoff frequency** on the gain-frequency plot is the frequency where the voltage gain has dropped by **3 dB (0.707)** from the dB voltage gain at the high frequencies. In this experiment, you will study the high-pass active filters in Figures 40-1 and 40-2. Notice that the high-pass filter RC circuit capacitors and resistors have been interchanged when compared to the low-pass filter RC circuits in Figures 39-1 and 39-2. The expected **cutoff frequency (f_C)** for the **one-pole high-pass filter** in Figure 40-1 and the **two-pole Sallen-Key high-pass Butterworth filter** in Figure 40-2 can be calculated from

$$f_c = \frac{1}{2\pi RC}$$

The two-pole Sallen-Key high-pass Butterworth filter in Figure 40-2 requires a **voltage gain of 1.586** to produce the Butterworth response. Therefore,

$$A_V = 1 + \frac{R_1}{R_2} = 1.586$$

and

$$\frac{R_1}{R_2} = 0.586$$

Ideally, a high-pass filter should pass all frequencies above the cutoff frequency (f_C). Because op-amps have a limited bandwidth, active filters have an upper-frequency limit on the high-pass response, making it a band-pass filter with a very wide bandwidth. For this reason, active filters are mostly used in **low frequency applications**, where the op-amp amplifier bandwidth is wide enough that it does not fall within the frequency range of the application.

Figure 40-1 First-Order (1-pole) High-Pass Active Filter

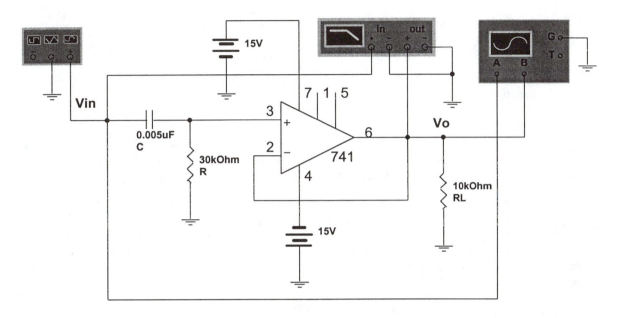

Figure 40-2 Second-Order (2-pole) Sallen-Key High-Pass Butterworth Filter

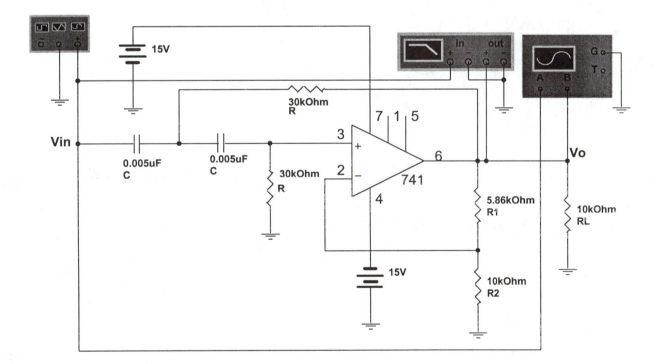

Procedure:

Step 1.   Open circuit file FIG40-1. Bring down the Bode plotter enlargement and make sure that the following settings are selected: *Magnitude*, Vertical (Log, F = 0 dB, I = −40 dB), Horizontal (Log, F = 200 kHz, I = 1 Hz). You will plot the frequency response curve for a first-order (one-pole) high-pass active filter as the frequency varies between 1.0 Hz and 200 kHz using the Bode plotter.

NOTE: If you are performing this experiment in a lab environment, you may not have a Bode plotter available. You will need to plot the frequency response curve by making dB voltage gain measurements at different frequencies between 1.0 Hz and 200 kHz and plotting the curve on semilog graph paper.

Step 2.   Run the simulation and notice that the frequency response curve has been plotted between the frequencies of 1.0 Hz and 200 kHz by the Bode plotter. Draw the curve plot in the space provided. Next, move the cursor to the flat part of the curve at a frequency of approximately 50 kHz and measure the voltage gain in dB. *Record the dB voltage gain on the curve plot.*

One-Pole High-Pass Filter

A_{dB}

f

Question: Is the frequency response curve that of a high-pass filter? **Explain.**

Step 3.   Calculate the actual voltage gain (A_V) from the dB voltage gain measured in Step 2.

Question: Was the voltage gain (A_V) on the flat part of the frequency response curve what you expected for the circuit in Figure 40-1? **Explain.**

Step 4.   Move the cursor as close as possible to a point on the curve that is 3 dB down from the dB voltage gain at the high frequencies measured in Step 2. Record the frequency (cutoff frequency, f_C) on the curve plot drawn in Step 2.

Step 5.   Calculate the expected cutoff frequency (f_C) based on the circuit component values.

Question: How did the calculated value for the cutoff frequency compare with the measured value recorded on the curve plot for the one-pole high-pass filter?

Step 6.   Move the cursor to a point on the curve where the frequency is as close as possible to one-tenth f_C. Record the new dB voltage gain and frequency (f_2) on the curve plot drawn in Step 2.

Question: Approximately how much did the dB voltage gain decrease for a one decade decrease in frequency for the one-pole high-pass filter?   Was this what you expected for a one-pole filter?

Step 7.   Click *Phase* on the Bode plotter to plot the phase curve. Change the vertical axis initial value (I) to 0° and the final value (F) to 90°. Run the simulation again. You are looking at the phase difference between the filter input and output waveshapes as a function of frequency. Draw the curve plot in the space provided.

One-Pole High-Pass Filter

ϕ

f

Step 8.   Move the cursor to the top of the curve plot (low frequency). Record the phase in degrees on the curve plot. Move the cursor as close as possible on the curve to a frequency that is equal to the value measured in Step 4 (cutoff frequency, f_C). Record the frequency and phase (ϕ) on the curve plot drawn in Step 7.

Question: Was the phase shift between input and output at the cutoff frequency what you expected for a one-pole high-pass filter? **Explain.**

Step 9.   Bring down the oscilloscope enlargement and make sure that the following settings are selected: Time base (Scale = 50 μs/Div, Xpos = 0, Y/T), Ch A (Scale = 1 V/Div, Ypos = 0, DC), Ch B (Scale = 500 mV/Div, Ypos = 0, DC), Trigger (Pos edge, Level = 0 V, Auto). Bring down the function generator enlargement and make sure that the following settings are selected: *Sine Wave*, Freq = 10 kHz, Ampl = 1 V, Offset = 0. Run the simulation and keep decreasing the function generator frequency. Notice what happens to the output sine wave magnitude and phase as you approach the cutoff frequency (f_C). Adjust the oscilloscope as needed. Record the peak-to-peak output voltage magnitude (V_o) and the phase (ϕ) between the input and output at f = 10 kHz and at f = f_C.

   V_o = _____ ϕ = _____   (at f = 10 kHz)

   V_o = _____ ϕ = _____   (at f = f_C)

Questions: Did the output magnitude do what you expected for a one-pole high-pass filter when you approached the cutoff frequency? Did it drop 3 dB (0.707) at the cutoff frequency?

Did the phase between the input and output change as you expected for a one-pole filter as you approached the cutoff frequency?

Step 10.   Return the oscilloscope and function generator to normal size to expose the Bode plotter. Click *Magnitude* on the plotter. Adjust the horizontal final frequency (F) on the Bode plotter to 50 MHz. Run the simulation and draw the curve plot in the space provided.

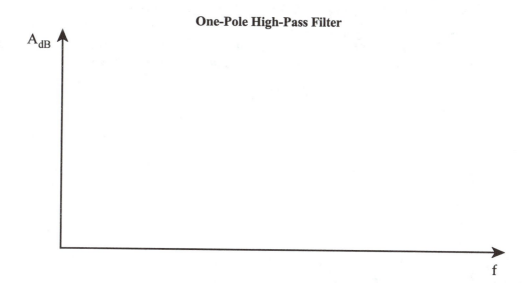

One-Pole High-Pass Filter

Step 11.   Measure the upper cutoff frequency (f_{C2}) following the procedure in Step 4. Record the value on the curve plot drawn in Step 10.

Question: Explain why the filter frequency response looked like a band-pass response when frequencies above 1 MHz were plotted in Step 10. *Hint:* What was the value of the unity-gain bandwidth, f_{unity}, for the 741 op-amp?

Step 12.   Open circuit file FIG40-2. Bring down the Bode plotter enlargement and make sure that the following settings are selected: *Magnitude*, Vertical (Log, F = 10 dB, I = –40 dB), Horizontal (Log, F = 200 kHz, I = 1 Hz). You will plot the frequency response curve for a second-order (two-pole) high-pass active filter as the frequency varies between 1.0 Hz and 200 kHz using the Bode plotter.

NOTE: If you are performing this experiment in a lab environment, you may not have a Bode plotter available. You will need to plot the frequency response curve by making dB voltage gain measurements a different frequencies between 1.0 Hz and 200 kHz and plotting the curve on semilog graph paper.

Step 13.   Run the simulation and notice that the frequency response curve has been plotted between the frequencies of 1.0 Hz and 200 kHz by the Bode plotter. Draw the curve plot in the space provided. Next, move the cursor to the flat part of the curve at a frequency of approximately 50 kHz and measure the voltage gain in dB. *Record the dB voltage gain on the curve plot.*

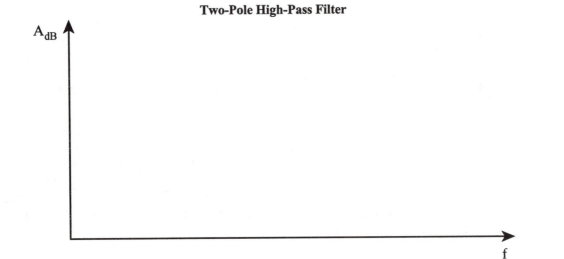

Two-Pole High-Pass Filter

Question: Is the frequency response curve that of a high-pass filter? **Explain.**

Step 14.   Calculate the actual voltage gain (A_V) from the dB voltage gain measured in Step 13.

Question: Was the voltage gain (A_V) on the flat part of the frequency response curve what you expected for the circuit in Figure 40-2? **Explain why.**

Step 15.   Move the cursor as close as possible to a point on the curve that is 3 dB down from the dB voltage gain at the high frequencies measured in Step 13. Record the frequency (cutoff frequency, f_C) on the curve plot drawn in Step 13.

Step 16.   Calculate the expected cutoff frequency (f_C) based on the circuit component values.

Question: How did the calculated value for the cutoff frequency compare with the measured value recorded on the curve plot for the two-pole high-pass filter?

Step 17.   Move the cursor to a point on the curve where the frequency is as close as possible to one-tenth f_C. Record the new dB voltage gain and frequency (f_2) on the curve plot drawn in Step 13.

Questions: Approximately how much did the dB voltage gain decrease for a one decade decrease in frequency for the two-pole high-pass filter?   Was this what you expected for a two-pole filter?

Step 18.   Click *Phase* on the Bode plotter to plot the phase curve. Change the vertical axis initial value (I) to 0° and the final value (F) to 180°. Run the simulation again. You are looking at the phase difference between the filter input and output waveshapes as a function of frequency. Draw the curve plot in the space provided.

Two-Pole High-Pass Filter

φ

f

Step 19.   Move the cursor to the top of the curve plot (low frequency). Record the phase in degrees on the curve plot. Move the cursor as close as possible on the curve to a frequency that is equal to the value measured in Step 15 (cutoff frequency, f_C). Record the frequency and phase (φ) on the curve plot drawn in Step 18.

Question: Was the phase shift between input and output at the cutoff frequency what you expected for a two-pole high-pass filter? **Explain.**

Step 20.   Bring down the oscilloscope enlargement and make sure that the following settings are selected: Time base (Scale = 50 μs/Div, Xpos = 0, Y/T), Ch A (Scale = 1 V/Div, Ypos = 0, DC), Ch B (Scale = 500 mV/Div, Ypos = 0, DC), Trigger (Pos edge, Level = 0 V, Auto). Bring down the function generator enlargement and make sure that the following settings are selected: *Sine Wave*, Freq = 10 kHz, Ampl = 1 V, Offset = 0. Run the simulation and keep decreasing the function generator frequency. Notice what happens to the output sine wave magnitude and phase as you approach the cutoff frequency (f_C). Adjust the oscilloscope as needed. Record the peak-to-peak output voltage magnitude (V_o) and the phase (φ) between the input and output at f = 10 kHz and at f = f_C.

V_o = _____   φ = _____   (at f = 10 kHz)

V_o = _____   φ = _____   (at f = f_C)

Questions: Did the output magnitude do what you expected for a two-pole high-pass filter when you approached the cutoff frequency in Step 20? Did it drop 3 dB (0.707) at the cutoff frequency?

Did the phase between the input and output change as you expected for a two-pole filter as you approached the cutoff frequency in Step 20?

Step 21.   Return the oscilloscope and function generator to normal size to expose the Bode plotter. Click *Magnitude* on the plotter and change the horizontal axis final setting (F) to 50 MHz. Run the simulation and draw the curve plot in the space provided.

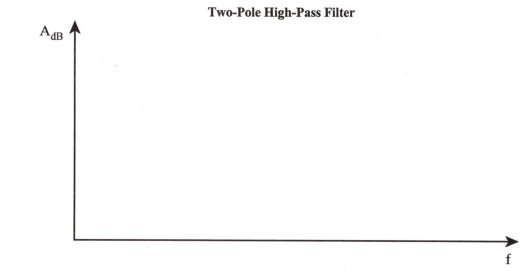

Two-Pole High-Pass Filter

Step 22.   Measure the upper cutoff frequency (f_{C2}) following the procedure in Step 15 and record the value on the curve plot drawn in Step 21.

Question: Explain why the filter frequency response looked like a band-pass response when frequencies above 1 MHz were plotted in Step 21. *Hint:* What was the value of the unity-gain bandwidth , f_{unity}, for the 741 op-amp?

Troubleshooting Problems

1. Open circuit file FIG40-3 and run the simulation. Locate the defective component and state the defect (short or open). You can use any instrument available and make any measurement desired.

    Defective component: _____   Defect: _____

2. Open circuit file FIG40-4 and run the simulation. Locate the defective component and state the defect (short or open). You can use any instrument available and make any measurement desired.

    Defective component: _____   Defect: _____

3. Open circuit file FIG40-5 and run the simulation. Locate the defective component and state the defect (short or open). You can use any instrument available and make any measurement desired.

    Defective component: _____   Defect: _____

Band-Pass Active Filters

Objectives:

1. Plot the frequency response curve for a wide-band two-pole band-pass active filter.
2. Plot the frequency response curve for a narrow-band two-pole band-pass active filter.
3. Determine the center frequency (f_o) of a band-pass active filter.
4. Determine the bandwidth of a band-pass active filter.
5. Determine the quality factor (Q) of a band-pass active filter.
6. Determine the roll-off in dB per decade for a two-pole band-pass active filter.
7. Plot the phase shift between the input and output for a two-pole band-pass active filter.

Materials:

Two 15 V dc voltage supplies
Two LM741 op-amps
One function generator
One dual-trace oscilloscope
Capacitors: two .001 µF, two .004 µF, two .01 µF
Resistors: one 1 kΩ, two 5.86 kΩ, three10 kΩ, four 30 kΩ, and one 100 kΩ.

Theory:

A **band-pass filter** passes all frequencies lying within a band of frequencies and rejects all other frequencies outside the band. The pass-band **low-cutoff frequency (f_{C1})** and the pass-band **high-cutoff frequency (f_{C2})** on the gain-frequency plot are the frequencies where the voltage gain has dropped 3 dB (0.707) from the maximum dB gain. The filter **bandwidth (BW)** is the difference between the high-cutoff frequency and the low-cutoff frequency. Therefore,

$$BW = f_{C2} - f_{C1}$$

The **center frequency (f_o)** of a band-pass filter is the geometric mean of the low-cutoff frequency (f_{C1}) and the high-cutoff frequency (f_{C2}). Therefore,

$$f_o = \sqrt{f_{C1}f_{C2}}$$

The **quality factor (Q)** of a band-pass filter is the ratio of the center frequency (f_o) and the bandwidth (BW), and is an indication of the selectivity of the filter. Therefore,

$$Q = \frac{f_o}{BW}$$

A higher value of Q means a narrower bandwidth and a more selective filter. A filter with a Q less than one is considered to be a **wide-band filter**, and a filter with a Q greater than one is considered to be a **narrow-band filter**.

One way to implement a band-pass filter is to cascade a low-pass and a high-pass filter. As long as the cutoff frequencies are sufficiently separated, the low-pass filter cutoff frequency will determine the low cutoff frequency of the pass-band and the high-pass filter cutoff frequency will determine the high cutoff frequency of the pass-band. Normally this arrangement is used for a wide-band filter (Q< 1) because the cutoff frequencies need to be sufficiently separated.

Figure 41-1 shows a **two-pole Sallen-Key low-pass Butterworth filter** and a **two-pole Sallen-Key high-pass Butterworth filter** cascaded to form a **two-pole band-pass filter**. The upper and lower cutoff frequencies are determined by the cutoff frequencies of the low and high-pass filters. The gain of a two-pole band-pass filter rolls off at a rate of **40 dB per decade**. The expected cutoff frequency for each two-pole Sallen-Key filter can be calculated from

$$f_c = \frac{1}{2\pi RC}$$

where $R = R_a$ or R_b and $C = C_a$ or C_b

The phase between the input and output for the two-pole high-pass filter goes from +180 degrees to 0 degrees over the frequency range. The phase between the input and output for the two-pole low-pass filter goes from 0 degrees to –180 degrees over the frequency range. Therefore, the **phase** for the combined band-pass filter should go from **+180 degrees to –180 degrees** over the frequency range, and should be at **0 degrees** at the cutoff frequency.

For a **narrow-band (Q > 1) band-pass filter**, a circuit such as the **multiple-feedback band-pass filter** shown in Figure 41-2 is required. Components R_1 and C_1 determine the low cutoff frequency, and R_2 and C_2 determine the high cutoff frequency. The center frequency (f_o) can be calculated from the component values using the equation

$$f_o = \frac{1}{2\pi\sqrt{R_1 R_2 C_1 C_2}} = \frac{1}{2\pi C\sqrt{R_1 R_2}}$$

where $C = C_1 = C_2$. The voltage gain (A_V) at the center frequency is calculated from

$$A_V = \frac{R_2}{2R_1}$$

and the quality factor (Q) is calculated from

$$Q = 0.5\sqrt{\frac{R_2}{R_1}}$$

Figure 41-1 Two-Pole Cascaded Butterworth Band-Pass Filter

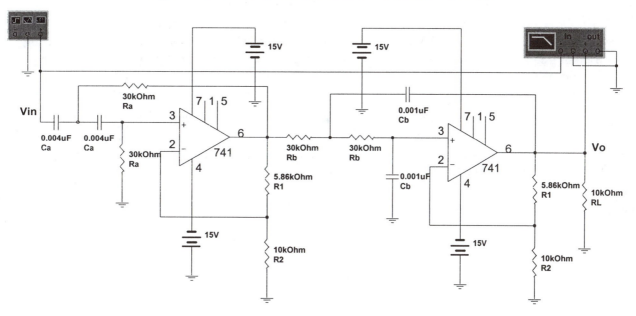

Figure 41-2 Multiple-Feedback Band-Pass Filter

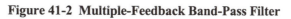
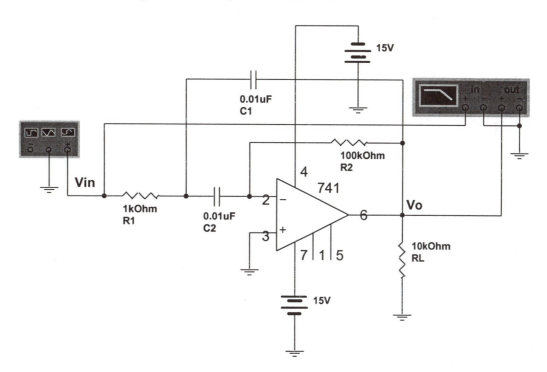

Procedure:

Step 1.   Open circuit file FIG41-1. Bring down the Bode plotter enlargement and make sure that the following settings are selected: *Magnitude*, Vertical (Log, F = 10 dB, I = –40 dB), Horizontal (Log, F = 60 kHz, I = 50 Hz). You will plot the frequency response curve for a second-order (two-pole) cascaded Butterworth band-pass filter as the frequency varies between 50 Hz and 60 kHz, using the Bode plotter.

> *NOTE:* If you are performing this experiment in a lab environment, you may not have a Bode plotter available. You will need to plot the frequency response curve by making dB voltage gain measurements at different frequencies between 50 Hz and 60 kHz using an oscilloscope and plotting the curve on semilog graph paper.

Step 2.   Run the simulation and notice that the frequency response curve for a band-pass filter has been plotted between the frequencies of 50 Hz and 60 kHz. Draw the curve plot in the space provided. Next, move the cursor to the center of the curve at its peak point. Measure the center frequency (f_o) and the voltage gain in dB. *Record the dB voltage gain and center frequency (f_o) on the curve plot.*

A_{dB}

f

Question: Is the frequency response curve that of a band-pass filter? **Explain why.**

Step 3.   Move the cursor as close as possible to a point on the left side of the curve that is 3 dB down from the dB voltage gain measured at the center frequency. Record the frequency (low-cutoff frequency, f_{C1}) on the curve plot. Next, move the cursor as close as possible to a point on the right side of the curve that is 3 dB down from the dB voltage gain measured at the center frequency. Record the frequency (high-cutoff frequency, f_{C2}) on the curve plot.

Step 4.   Based on the values of f_{C1} and f_{C2} measured in Step 3, determine the bandwidth (BW) of the band-pass filter.

Step 5.   Calculate the expected cutoff frequencies (f_{C1} and f_{C2}) for each cascaded active filter based on the circuit component values.

Question: How did the calculated values for the cutoff frequencies (f_{C1} and f_{C2}) compare with the measured values recorded on the curve plot for the two-pole band-pass filter?

Step 6.   Based on the values calculated for f_{C1} and f_{C2} in Step 5, calculate the expected center frequency (f_o).

Question: How did the calculated value of the center frequency (f_o) compare with the measured value on the curve plot for the two-pole band-pass filter?

Step 7.   Based on the measured center frequency (f_o) and bandwidth (BW), calculate the quality factor (Q) for the band-pass filter.

Question: Based on the quality factor (Q) for this band-pass filter, would you consider this to be a wide-band or a narrow-band filter? **Explain why.**

Step 8.   Move the cursor to a point on the curve where the frequency is as close as possible to ten times f_{C2}. Record the new dB voltage gain and frequency (f_3) on the curve plot.

Question: Approximately how much did the dB voltage gain decrease for a one decade increase in frequency for this two-pole band-pass filter?   Was this what you expected for a two-pole filter?

Step 9.   Click *Phase* on the Bode plotter to plot the phase curve. Change the vertical axis initial value (I) to $-180°$ and the final value (F) to $180°$. Run the simulation again. You are looking at the phase difference between the filter input and output waveshapes as a function of frequency. Draw the curve plot in the space provided.

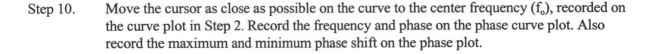

Step 10.   Move the cursor as close as possible on the curve to the center frequency (f_o), recorded on the curve plot in Step 2. Record the frequency and phase on the phase curve plot. Also record the maximum and minimum phase shift on the phase plot.

Questions: Was the phase between input and output at the center frequency (f_o) what you expected for a two-pole band-pass filter?

Did the phase between the input and output change over the frequency range as you expected for a two-pole band-pass filter?

Step 11.   Open circuit file FIG41-2. Bring down the Bode plotter enlargement and make sure that the following settings are selected: *Magnitude*, Vertical (Log, F = 40 dB, I = 10 dB), Horizontal (Log, F = 10 kHz, I = 100 Hz). You will plot the frequency response curve for a multiple-feedback band-pass filter as the frequency varies between 100 Hz and 10 kHz, using the Bode plotter.

NOTE: If you are performing this experiment in a lab environment, you may not have a Bode plotter available. You will need to plot the frequency response curve by making dB voltage gain measurements at different frequencies between 100 Hz and 10 kHz using an oscilloscope and plotting the curve on semilog graph paper.

Step 12.   Run the simulation and notice that the frequency response curve for a band-pass filter has been plotted between the frequencies of 100 Hz and 10 kHz. Draw the curve plot in the space provided. Next, move the cursor to the center of the curve at its peak point. Measure the center frequency (f_o) and the voltage gain in dB. *Record the dB voltage gain and center frequency (f_o) on the curve plot.*

A_{dB}

f

Question: Is the frequency response curve that of a band-pass filter? **Explain why.**

Step 13.   Based on the dB voltage gain at the center frequency measured in Step 12, calculate the actual voltage gain (A_V).

Step 14.   Based on the circuit component values, calculate the expected voltage gain at the center frequency (f_o).

Question: How did the measured voltage gain at the center frequency compare with the voltage gain calculated from the circuit values?

Step 15.   Move the cursor as close as possible to a point on the left side of the curve that is 3 dB down from the peak dB voltage gain. Record the frequency (low-cutoff frequency, f_{C1}) on the curve plot. Next, move the cursor as close as possible to a point on the right side of the curve that is 3 dB down from the peak dB voltage gain. Record the frequency (high-cutoff frequency, f_{C2}) on the curve plot.

Step 16.   Based on the values of f_{C1} and f_{C2} measured in Step 15, calculate the bandwidth (BW) of the band-pass filter.

Step 17.   Based on the circuit component values, calculate the expected center frequency (f_o).

Question: How did the calculated value of the center frequency compare with the value measured in Step 12?

Step 18.   Based on the center frequency (f_o) and the bandwidth (BW), calculate the quality factor (Q) of this band-pass filter.

Question: Based on the value of Q, would you consider this a narrow-band filter or a wide-band filter? **Explain why.**

Step 19.   Based on the component values, calculate the expected quality factor (Q) of this band-pass filter.

Question: How did your calculated value of Q based on the component values compare with the value of Q determined in Step 18 from the measured f_o and BW?

Troubleshooting Problems

1. Open circuit file FIG41-3. Run the simulation. Locate the defective component and state the defect (short or open). You can use any instrument available and make any measurement desired.

    Defective component: _____   Defect: _____

2. Open circuit file FIG41-4. Run the simulation. Locate the defective component and state the defect (short or open). You can use any instrument available and make any measurement desired.

    Defective component: _____   Defect: _____

3. Open circuit file FIG41-5. Run the simulation. Locate the defective component and state the defect (short or open). You can use any instrument available and make any measurement desired.

    Defective component: _____   Defect: _____

4. Open circuit file FIG41-6. Run the simulation. Locate the defective component and state the defect (short or open). You can use any instrument available and make any measurement desired.

    Defective component: _____   Defect: _____

Band-Stop (Notch) Active Filters

Objectives:

1. Plot the frequency response curve for a band-stop (notch) active filter.
2. Determine the center frequency (f_o) of a notch active filter.
3. Determine the bandwidth of a notch active filter.
4. Determine the quality factor (Q) of a notch active filter.
5. Determine the passband voltage gain of a notch active filter.

Materials:

Two dc voltage supplies
One LM741 op-amp
One function generator
One dual-trace oscilloscope
Capacitors: two 0.05 µF, one 0.1 µF
Resistors: two10 kΩ, one 13 kΩ, one 27 kΩ, two 54 kΩ

Theory:

A **band-stop** filter rejects a band of frequencies and passes all other frequencies outside the band, and is often referred to as a band-reject or **notch** filter. The **low-cutoff frequency (f_{C1})** and the **high-cutoff frequency (f_{C2})** on the gain-frequency plot are the frequencies where the voltage gain has dropped **3 dB (0.707)** from the passband dB voltage gain. The filter **bandwidth (BW)** is the difference between the high-cutoff frequency and the low-cutoff frequency. Therefore,

$$BW = f_{C2} - f_{C1}$$

The **center frequency (f_o)** of a notch filter is the geometric mean of the low-cutoff frequency (f_{C1}) and the high-cutoff frequency (f_{C2}). Therefore,

$$f_o = \sqrt{f_{C1}f_{C2}}$$

The **quality factor (Q)** of a notch filter is the ratio of the center frequency (f_o) and the bandwidth (BW), and is an indication of the **selectivity** of the filter. Therefore,

$$Q = \frac{f_o}{BW}$$

A higher value of Q means a narrower bandwidth and a more selective filter. A filter with a Q less than one is considered to be a **wide-band filter**, and a filter with a Q greater than one is considered to be a **narrow-band filter**.

Figure 42-1 shows a **Sallen-Key second-order (two-pole) notch filter**. The expected center frequency (f_o) can be calculated from

$$f_o = \frac{1}{2\pi RC}$$

At this frequency (f_o), the feedback signal returns with the correct amplitude and phase to attenuate the input. This causes the output to be attenuated at the center frequency.

The notch filter in Figure 42-1 has a **passband voltage gain (A_V)** that can be calculated from

$$A_V = \frac{R_2}{R_1} + 1$$

and a quality factor (Q) that can be calculated from

$$Q = \frac{0.5}{2 - A_V}$$

The voltage gain of a Sallen-Key notch filter must be less than 2 and the circuit Q must be much less than 10 to avoid oscillation.

Figure 42-1 Two-Pole Sallen-Key Notch Filter

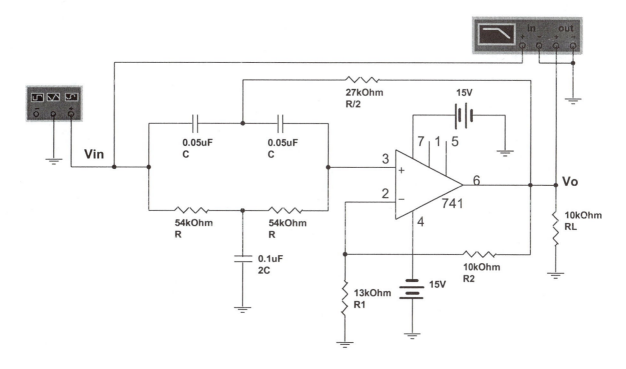

Procedure:

Step 1.   Open circuit file FIG42-1. Bring down the Bode plotter enlargement and make sure that the following settings are selected: *Magnitude*, Vertical (Log, F = 10 dB, I = –20 dB), Horizontal (Log, F = 500 Hz, I = 2 Hz). You will plot the frequency response curve for a second-order (two-pole) Sallen-Key notch filter as the frequency varies between 2 Hz and 500 Hz using the Bode plotter.

NOTE: If you are performing this experiment in a lab environment, you may not have a Bode plotter available. You will need to plot the frequency response curve by making dB voltage gain measurements at different frequencies between 2 Hz and 500 Hz using an oscilloscope and plotting the curve on semilog graph paper.

Step 2.   Run the simulation and notice that the frequency response curve for a notch filter has been plotted between the frequencies of 2 Hz and 500 Hz. Draw the curve plot in the space provided. (See next page.) Next, move the cursor to the center of the curve at its lowest point. Measure the center frequency (f_o) and record it on the curve plot. Next, move the cursor to the flat part of the curve in the pass-band. Measure the voltage gain in dB and record the dB voltage gain on the curve plot.

A_{dB}

f

Question: Is the frequency response curve that of a notch filter? **Explain why.**

Step 3.   Based on the measured passband dB voltage gain in Step 2, calculate the actual voltage gain (A_V).

Step 4.   Based on the circuit component values, calculate the expected passband voltage gain.

Question: How did the measured passband voltage gain compare with the voltage gain calculated from the circuit values?

Step 5.   Move the cursor as close as possible to a point on the left side of the curve that is 3 dB down from the dB voltage gain on the flat part of the curve plot. Record the frequency (low-cutoff frequency, f_{C1}) on the curve plot. Next, move the cursor as close as possible to a point on the right side of the curve that is 3 dB down from the dB voltage gain on the flat part of the curve plot. Record the frequency (high-cutoff frequency, f_{C2}) on the curve plot.

Step 6.　　　Based on the values of f_{C1} and f_{C2} measured in Step 5, calculate the bandwidth (BW) of the notch filter.

Step 7.　　　Based on the circuit component values, calculate the expected center frequency (f_o).

Question: How did the calculated value of the center frequency compare with the value measured on the curve plot?

Step 8.　　　Based on the center frequency (f_o) on the curve plot and the bandwidth (BW) determined in Step 6, calculate the quality factor (Q) of the notch filter.

Question: Based on the value of Q, would you consider this a narrow-band filter or a wide-band filter? **Explain why.**

Step 9.　　　Based on the pass-band voltage gain (A_V), calculate the expected quality factor (Q) of the notch filter.

Question: How did your calculated value of Q based on the pass-band voltage gain compare with the value of Q determined in Step 8 from the measured f_o and BW?

Troubleshooting Problems

1.   Open circuit file FIG42-2. Run the simulation. Locate the defective component and state the defect (short or open). You can use any instrument available and make any measurement desired.

        Defective component: _____   Defect: _____

2.   Open circuit file FIG42-3. Run the simulation. Locate the defective component and state the defect (short or open). You can use any instrument available and make any measurement desired.

        Defective component: _____   Defect: _____

3.   Open circuit file FIG42-4. Run the simulation. Locate the defective component and state the defect (short or open). You can use any instrument available and make any measurement desired.

        Defective component: _____   Defect: _____

VII

Oscillators

In the three experiments in Part VII, you will analyze a Wien-bridge oscillator, a Colpitts oscillator, and an astable multivibrator using a 555 timer. The Wien-bridge and Colpitts oscillators will generate a sine wave, and the astable multivibrator will generate a square wave. The analysis of the Wien-bridge oscillator and the Colpitts oscillator will show how the sine wave oscillation builds up until the gain-feedback product (AB) is equal to 1 at the resonant frequency of the feedback network. If you are performing these experiments in a lab environment, you will not be able to observe the oscillation buildup because it occurs in such a short time period. This is one of the advantages of using a computer circuit simulator. With a circuit simulator, you can slow down the transient time behavior and observe it on the screen.

The circuits for the experiments in Part VII can be found on the enclosed disk in the OSC subdirectory.

43

Name_____

Date_____

Wien-Bridge Oscillator

Objectives:

1. Observe how the output of an oscillator builds up until the gain-feedback product (AB) is equal to 1 at the resonant frequency of the feedback network.
2. Measure the frequency of oscillation of a Wien-bridge oscillator.
3. Learn what determines the frequency of oscillation of a Wien-bridge oscillator.
4. Learn what determines the peak-to-peak output voltage of a Wien-bridge oscillator.
5. Observe what happens to the oscillator output when output saturation is reached.

Materials:

Two dc voltage supplies
One 741 op-amp
Two 1N4001 diodes
One dual-trace oscilloscope
Capacitors (two each):   0.01 μF, 0.02 μF
Resistors: two 15 kΩ, two 20 kΩ, one 25 kΩ, one 27 kΩ, one 100 kΩ

Theory:

An **oscillator** is a circuit that generates an output signal, such as a sine wave, square wave, or a sawtooth wave, without an input signal. With the exception of the relaxation oscillator, an oscillator is an **amplifier with positive feedback**, where a portion of the output signal is fed back to the input, in phase with the input. If the feedback voltage is large enough, the amplifier will sustain a continuous output signal without any input signal. The oscillator will oscillate at the particular frequency that allows the output to be fed back to the input in phase with the input. This frequency is determined by the **resonant frequency** of the **feedback filter network**. In order for oscillation to be sustained, the total voltage gain around the closed feedback loop must equal unity (1). The voltage gain around the closed feedback loop is the product of the **amplifier voltage gain (A)** and the **feedback network attenuation (B)**. Therefore, for oscillation to be sustained,

   AB = 1

In order for oscillation to begin, the **closed-loop gain (AB)** must be greater than unity (1) when the oscillator is started. As the output voltage builds up, the amplifier gain will go down until the closed-loop gain (AB) is equal to unity (1). If the amplifier does not go into saturation before AB = 1, then a sine wave is produced at the output. If the amplifier goes into saturation before AB = 1, then a square wave is produced at the output. In order to produce a sine wave output, the closed-loop gain (AB) should be only slightly above unity (1) when the oscillator is started.

The **Wien-bridge oscillator** is the standard oscillator for low to moderate frequencies, in the range of 5 Hz to 1 MHz. In this experiment, you will study the Wien-bridge oscillator shown in Figure 43-1. The RC network feeds back some of the amplifier output to the input and forms the Wien-bridge feedback network (B). The 741 op-amp with resistors R_1, R_2, the IN4001 diodes, and the 15 kΩ resistor forms the amplifier (A). The transient analysis will show how the oscillation builds up to a steady-state sine wave over a period of time, until AB is equal to unity (1).

One of the requirements for sustained oscillation is having a **360° (0°) phase shift** in the feedback network at the resonant frequency, which does occur in a Wien-bridge filter. The other requirement for sustained oscillation is having the **gain-feedback product (AB) equal unity (1)** at the resonant frequency of the feedback network. Because a Wien-bridge filter attenuation (B) is 1/3 at the resonant frequency, the amplifier voltage gain (A) must be equal to 3 for oscillation to be sustained. This means that the voltage gain must be slightly above 3 when the oscillator is started. Resistor R_1, the parallel combination of the diode resistance and the 15 kΩ resistor, and resistor R_2 determine the amplifier voltage gain (A). When the simulation is started, there will be no diode current and the diode resistance will be high. This will make the amplifier voltage gain (A) slightly greater than 3. The current in the diodes will increase as the output builds up, thus lowering the diode resistance until the amplifier voltage gain is reduced to 3. When this happens, the closed loop gain (AB) will equal unity (AB = 3 × 1/3 = 1) and the output sine wave will be at a steady-state amplitude at the resonant frequency of the feedback network. The steady-state output amplitude will depend on the value of R_1, as will be demonstrated. Changing the value of R_1 will change the diode current required to bring the op-amp amplifier voltage gain to 3. This will require a different peak-to-peak output voltage. If R_1 is too high, the amplifier will be driven into saturation before reaching a voltage gain of 3, and the output will be a square wave instead of a sine wave.

The resonant frequency of the Wien-bridge filter network will determine the **frequency of oscillation (f)** of the oscillator and is calculated from

$$f = \frac{1}{2\pi RC}$$

The voltage gain of the amplifier (A) is equal to the ac peak-to-peak amplifier output voltage (V_o) divided by the ac peak-to-peak amplifier input voltage (V_{in}). Therefore,

$$A = \frac{V_o}{V_{in}}$$

The frequency of a sine wave (f) is the inverse of the time period of one cycle (T). Therefore,

$$f = \frac{1}{T}$$

Figure 43-1 Wien-Bridge Oscillator

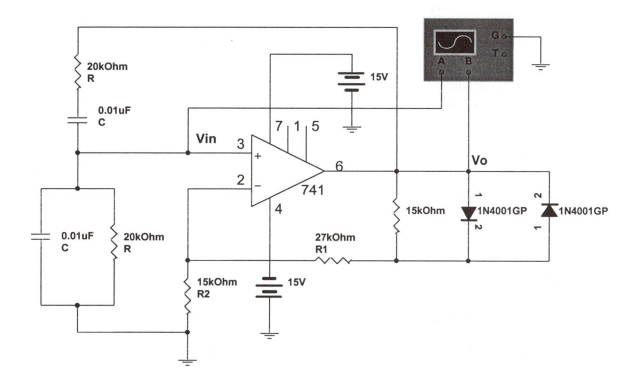

Procedure:

Step 1.   Open circuit file FIG43-1. Bring down the oscilloscope enlargement and make sure that the following settings are selected: Time base (Scale = 500 μs/Div, Xpos = 0, Y/T), Ch A (Scale = 5 V/Div, Ypos = 0, DC), Ch B (Scale = 5 V/Div, Ypos = 0, DC), Trigger (Pos edge, Level = 0 V, Auto). In this experiment you will observe how the output of an oscillator builds up until the gain-feedback product (AB) is equal to 1 at the resonant frequency of the feedback network.

NOTE: If you are performing this experiment in a lab environment, you may not be able to observe the oscillation build up because it occurs in such a short time period. With a circuit simulator, you can slow down the transient time behavior and observe it on the screen.

Step 2.   Run the simulation and notice the sine wave oscillation build up (after 0.025 seconds transient time) to a constant peak-to-peak voltage. Once the oscillator output reaches a constant peak-to-peak sine wave, click *Pause* to pause the display on the oscilloscope screen so that you can make some measurements. If you wish to continue the simulation, click *Resume*.

Question: Explain why the oscillator output builds up from zero to a constant peak-to-peak sine wave over a period of time.

Step 3.   Measure the time period (T) for one cycle of the output sine wave, the op-amp peak-to-peak output voltage (V_o), and the op-amp peak-to-peak input voltage (V_{in}). Record your measurements.

T = _____ V_o = _____ V_{in} = _____

Step 4.   Based on the time period (T) measured in Step 3, determine the frequency of oscillation.

Step 5.   Based on the Wien-bridge filter component values, calculate the expected oscillator output frequency.

Question: How did the measured frequency of oscillation compare with the calculated value?

Step 6.   Based on the op-amp peak-to-peak output and input voltages measured in Step 3, calculate the amplifier voltage gain (A).

Question: Was the amplifier gain what you expected for a Wien-bridge oscillator? **Explain why.**

Step 7.   Change the value of capacitor C (in both places) to 0.02 μF. Run the simulation again. (It will take approximately 0.04 seconds transient time for oscillation to begin.) Once the oscillator output reaches a constant peak-to-peak sine wave on the oscilloscope, click *Pause* to pause the display. Measure the time period (T) for one cycle of the sine wave, the op-amp peak-to-peak output voltage (V_o), and the op-amp peak-to-peak input voltage (V_{in}). Record your measurements.

          T = _____ V_o = _____ V_{in} = _____

Step 8.   Based on the time period (T) measured in Step 7, determine the frequency of oscillation.

Questions: What happened to the oscillator frequency when the value of capacitor C was changed? Did it affect the peak-to-peak output voltage? **Explain why.**

What determines the frequency of oscillation in the Wien-bridge oscillator?

Step 9.   Based on the new value for C, calculate the expected oscillator output frequency.

Question: How did the measured frequency of oscillation compare with the calculated value?

Step 10.   Change resistor R_1 to 25 kΩ. Run the simulation again. (It will take 0.05 seconds for oscillation to begin.) Once the oscillator reaches a constant peak-to-peak sine wave output voltage on the oscilloscope, click *Pause* to pause the output. Measure the time period (T) for one cycle of the sine wave, the op-amp peak-to-peak output voltage (V_o), and the op-amp peak-to-peak input voltage (V_{in}). Adjust the oscilloscope as needed. Record your measurements.

        T = _____ V_o = _____ V_{in} = _____

Questions: What happened to the peak-to-peak output voltage when you reduced R_1? Did changing R_1 affect the frequency?

Explain why the peak-to-peak output voltage changed when resistor R_1 was changed.

Step 11.   Change resistor R_1 to 100 kΩ. Run the simulation again. Once the oscillator reaches a constant peak-to-peak output voltage, click *Pause* to pause the display. Adjust the oscilloscope as needed. Draw the output waveshape in the space provided. *Note the peak output voltage on the curve plot.*

Questions: What happened to the output waveshape when the value of R_1 was increased to 100 kΩ? **Explain the reason for this result.**

Explain why the gain-feedback product (AB) must equal 1 at the resonant frequency of the positive feedback network in order for oscillation to occur.

Colpitts Oscillator

Objectives:

1. Observe how the output of an oscillator builds up until the gain-feedback product (AB) is equal to unity (1) at the resonant frequency of the feedback network.
2. Measure the frequency of oscillation of a Colpitts oscillator.
3. Learn what determines the frequency of oscillation of a Colpitts oscillator.
4. Learn how the LC filter capacitance ratio affects the amplifier voltage gain needed to sustain oscillation in a Colpitts oscillator.
5. Learn how the LC filter capacitance ratio affects the oscillator peak-to-peak output voltage in a Colpitts oscillator.
6. Explain how positive feedback is achieved in a Colpitts oscillator.

Materials:

One dc power supply
One dual-trace oscilloscope
One 2N3904 bipolar transistor
Capacitors: one 0.001 μF, one 0.01 μF, three 0.1 μF, one 500 pF
Inductors (one each): 5μH, 10 μH
Resistors: three 2 kΩ, one 15 kΩ, one 200 kΩ

Theory:

An **oscillator** is an **amplifier** with **positive feedback** and will oscillate at a frequency that allows the output to be fed back to the input in phase with the input. This frequency is determined by the **resonant frequency** of the **feedback network**. In order for oscillation to be sustained, the total **voltage gain around the closed loop (AB)** must equal **unity (1)** at the oscillation frequency. Therefore,

$$AB = 1$$

where B is the **oscillator feedback ratio** and A is the **amplifier voltage gain**. In order for oscillation to begin, the closed loop gain (AB) must be greater than unity (1) when the oscillator is started. As the output increases in amplitude, the voltage gain of the amplifier (A) drops until the closed loop gain (AB) is equal to unity (1). If the closed-loop gain is too large when the oscillator is started, the amplifier will be driven into **saturation** before the gain of the amplifier drops low enough to reduce the closed-loop gain to unity (1). This will cause the oscillator output to be a **square wave** instead of a sine wave.

At frequencies below 1 MHz, an oscillator with an RC feedback network, such as the Wien-bridge oscillator, is suitable. At frequencies above 1 MHz, an oscillator with an LC feedback network is normally used. Also, because of the frequency limitations of most op-amps, transistor amplifiers are mostly used as the gain elements in high-frequency LC oscillators. A popular LC oscillator is the **Colpitts oscillator**, shown in Figure 44-1. This oscillator uses the 180° phase shift in the **tuned LC circuit** in the feedback network and the 180° phase shift in the **transistor amplifier** to provide the required 360° phase shift. In this experiment, you will study the Colpitts oscillator shown in Figure 44-1.

The simulation will show how the sine wave oscillation builds up to a steady state over a period of time at the resonant frequency of the LC network until the closed-loop gain (AB) is equal to unity (1), which is the principal requirement for sustaining oscillation. Because the oscillator feedback ratio (B) is C_1/C_2 at the resonant frequency, the amplifier voltage gain (A) needed to sustain oscillation can be calculated from

$$AB = 1$$

$$A\left(\frac{C_1}{C_2}\right) = 1$$

$$A = \frac{C_2}{C_1}$$

The **oscillator peak-to-peak output voltage** is affected by the amplifier voltage gain (A). A higher voltage gain (A) causes the oscillator to require a lower output voltage to cause the closed-loop gain (AB) to reach unity (1). Therefore, the ratio of the capacitance values of capacitors C_1 and C_2 affects the oscillator peak-to-peak output voltage.

The resonant frequency of the LC filter determines the **frequency of oscillation (f)** of the oscillator and is approximated from

$$f = \frac{1}{2\pi\sqrt{LC_T}}$$

where $C_T = C_1C_2/(C_1 + C_2)$. This equation is accurate only if the LC circuit has a high Q.

The **frequency** of a sine wave (f) is the inverse of the **time period** for one cycle (T). Therefore,

$$f = \frac{1}{T}$$

The voltage gain of the amplifier (A) is equal to the ac peak-to-peak output voltage (V_o) divided by the ac peak-to-peak input voltage (V_{in}). Therefore,

$$A = \frac{V_o}{V_{in}}$$

Figure 44-1   Colpitts Oscillator

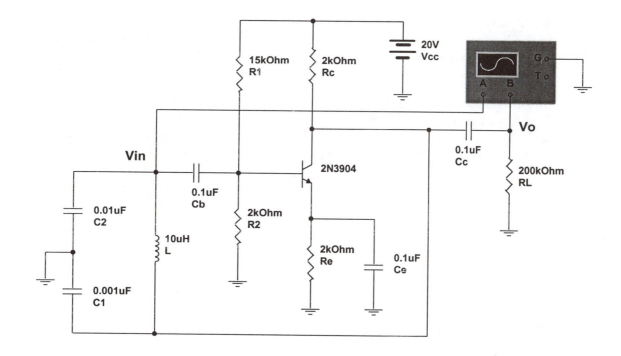

Procedure:

Step 1.   Open circuit file FIG44-1. Bring down the oscilloscope enlargement and make sure that the following settings are selected: Time base (Scale = 200 ns/Div, Xpos = 0, Y/T), Ch A (Scale = 500 mV/Div, Ypos = 0, AC), Ch B (Scale = 2 V/Div, Ypos = 0, AC), Trigger (Pos edge, Level = 0 V, Auto). In this experiment you will observe how the output of an oscillator builds up until the gain-feedback product (AB) is equal to unity (1) at the resonant frequency of the feedback network.

NOTE: If you are performing this experiment in a lab environment, you will not be able to observe the oscillation buildup because it occurs in such a short time period. With a circuit simulator, you can slow down the transient time behavior and observe the oscillation buildup on the computer screen.

Step 2.   Run the simulation and notice the sine wave oscillation build up after a 0.02 ms transient time. Once the oscillator output (blue curve plot) reaches a constant peak-to-peak sine wave, click *Pause* to pause the display on the oscilloscope screen so that you can make some measurements. If you wish to continue the simulation, click *Resume*.

Question: Explain why the oscillator output builds up from zero to a constant peak-to-peak sine wave over a period of time.

Step 3.   Measure the time period (T) for one cycle of the sine wave, the amplifier peak-to-peak output voltage (blue), the amplifier peak-to-peak input voltage (red), and the phase difference between the amplifier input and output. Record your measurements.

T = _____ V_o = _____ V_{in} = _____

Phase difference = _____ degrees

Question: Was the phase difference between the amplifier input and output what you expected? **Explain how positive feedback is achieved in a Colpitts oscillator.**

Step 4.   Based on the time period (T) measured in Step 3, determine the frequency of oscillation (f).

Step 5.   Based on the LC filter component values, calculate the expected oscillator output frequency.

Question: How did the measured frequency of oscillation compare with the calculated value?

Step 6.   Based on the amplifier peak-to-peak output and input voltages measured in Step 3, calculate the amplifier voltage gain (A).

Step 7.   Based on the values of LC network capacitors C_1 and C_2, calculate the amplifier voltage gain (A) needed to sustain oscillation.

Question: How did the measured amplifier voltage gain compare with the value needed to sustain oscillation? **Explain.**

Step 8.   Change the value of inductor L in the LC network to 5 µH. Run the simulation again. Once the oscillator output (blue curve plot) reaches a constant peak-to-peak sine wave, click *Pause* to pause the display. Measure the time period (T) for one cycle of the sine wave, the amplifier ac peak-to-peak output voltage (blue), and the amplifier ac peak-to-peak input voltage (red). Record your measurements.

$T =$ _____ $V_o =$ _____ $V_{in} =$ _____

Step 9.   Based on the time period (T) measured in Step 8, determine the frequency of oscillation (f).

Questions: What happened to the oscillator frequency when the value of inductor L was changed? Did it affect the peak-to-peak output voltage? **Explain.**

Step 10.   Based on the new value for L, calculate the expected oscillator output frequency (f).

Question: How did the measured frequency of oscillation compare with the calculated value?

Step 11.   Change the value of capacitor C_1 in the LC network to 500 pF. Run the simulation again. Once the oscillator reaches a constant peak-to-peak sine wave output (blue curve plot) voltage, click *Pause* to pause the display. Measure the time period (T) for one cycle of the sine wave, the amplifier ac peak-to-peak output voltage (blue), and the amplifier ac peak-to-peak input voltage (red). Record your measurements.

$T =$ _____ $V_o =$ _____ $V_{in} =$ _____

Step 12.   Based on the time period (T) measured in Step 11, determine the frequency of oscillation.

Question: Did the frequency change when C_1 was changed? **Explain.**

Step 13.   Based on the new value of C_1, calculate the expected oscillator output frequency (f).

Question: How did the measured frequency of oscillation compare with the calculated value?

Step 14.   Based on the amplifier peak-to-peak output and input voltages measured in Step 11, calculate the amplifier voltage gain (A).

Questions: Explain why the amplifier voltage gain changed when Capacitor C_1 was changed in Step 11. Was the frequency affected? **Explain why.**

Did the oscillator peak-to-peak output voltage change? **If so, why?**

Step 15.   Based on the values of LC network capacitors C_1 and C_2, calculate the amplifier voltage gain
                (A) needed to sustain oscillation.

Questions: How did the measured amplifier voltage gain compare with the value needed to sustain
oscillation? **Explain.**

Explain why the gain-feedback product (AB) must equal 1 at the resonant frequency of the feedback
network in order for oscillation to occur.

What determines the frequency of oscillation in the Colpitts oscillator?

45

Astable Multivibrator

Objectives:

1. Learn how to connect a 555 timer as an astable multivibrator (pulse generator).
2. Observe the output of a 555 timer wired as an astable multivibrator (pulse generator).
3. Measure the frequency of oscillation of a 555 timer connected as an astable multivibrator.
4. Learn what determines the frequency of oscillation of a 555 timer connected as an astable multivibrator.
5. Learn how to use the 555 timer as a voltage-controlled oscillator.

Materials:

One dc voltage supply
One dual-trace oscilloscope
One 555 timer
Capacitors: two 0.01 μF, one 0.02 μF
Resistors: one 1 kΩ, two 20 kΩ, one 1 kΩ potentiometer

Theory:

The **555 timer** is an **integrated circuit (IC)** that is commonly used as an **astable multivibrator (pulse generator)**, a **monostable multivibrator (one shot)** or a **voltage-controlled oscillator**. The 555 timer consists of two comparators, a resistive voltage divider, a flip-flop, a discharge transistor, and a buffer amplifier at the output. It is a two-state device whose output can be either high or low. The state of the output can be controlled by input signals and time-delay elements connected externally to the 555 timer.

In this experiment, the 555 timer will be wired as an astable multivibrator (pulse generator), as shown in Figure 45-1. The external components R_1, R_2, and C form the **timing circuit** that determines the **pulse frequency**. The 0.01 μF capacitor connected to the **control (CON) input** is for decoupling, and has no effect on the operation. When power is turned on, the capacitor (C) charges through resistors R_1 and R_2 until the **upper trip point (UTP) voltage** ($2V_{CC}/3$) is reached on the **trigger terminal (TRI)**. When this happens, the output is triggered, causing the **output voltage (OUT)** to go down. This also causes the internal discharge transistor to turn on and short the **discharge terminal (DIS)** to ground, discharging the capacitor through resistor R_2. When the voltage across the capacitor drops to the **lower trip point (LTP)** voltage ($V_{CC}/3$) on the **threshold input (THR)**, the discharging transistor turns off, causing the capacitor to start charging again through resistors R_1 and R_2 to begin the cycle again. At the same time, the output is triggered again, causing the output voltage (OUT) to go up. The result is a pulse output with a **duty cycle** that depends on the resistance values of R_1 and R_2. The time that it takes the capacitor to charge to the trigger voltage during the charging cycle determines the **time that the output is high (t_H)**. The value of t_H

can be calculated from

$$t_H = 0.693 \, (R_1 + R_2) \, C$$

The time that it takes the capacitor to discharge to the threshold voltage during the discharging cycle determines the **time that the output is low (t_L)**. The value of t_L can be calculated from

$$t_L = 0.693 \, R_2 \, C$$

The **time period (T)** for one complete cycle of the output square wave is equal to the sum of t_H and t_L. Therefore,

$$T = t_H + t_L$$

The **frequency of oscillation (f)** is equal to the inverse of the **time period (T)** for one cycle. Therefore,

$$f = \frac{1}{T}$$

The **duty cycle (D)** is equal to the ratio of the output high time (t_H) and the total time period (T) for one cycle as a percentage. Therefore,

$$D = \frac{t_H}{T} \times 100\%$$

In order to achieve a **50% duty cycle** (same up time and down time), t_H must be equal to one-half the total time period (T). This can only be achieved if R_1 is zero, which is not allowed on the 555 timer. Therefore, with the 555 timer circuit shown in Figure 45-1, a 50% duty cycle can only be approached if R_2 is much greater than R_1.

The 555 timer can be used as a **voltage-controlled oscillator (VCO)** by using the same external connections as for the astable multivibrator, except that a variable control voltage is applied to the **CON input** in place of the 0.01 µF capacitor, as shown in Figure 45-2. When the control voltage is varied, the oscillator frequency will also vary. An increase in the control voltage increases the threshold voltage, increasing the capacitor (C) charge and discharge time, which causes the frequency to decrease. A decrease in the control voltage decreases the threshold voltage, decreasing the capacitor (C) charge and discharge time, which causes the frequency to increase.

Figure 45-1 Astable Multivibrator

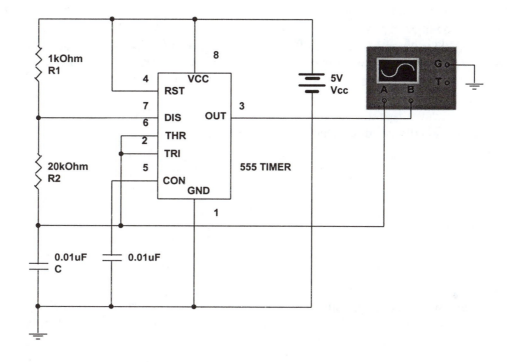

Figure 45-2 Voltage-Controlled Oscillator

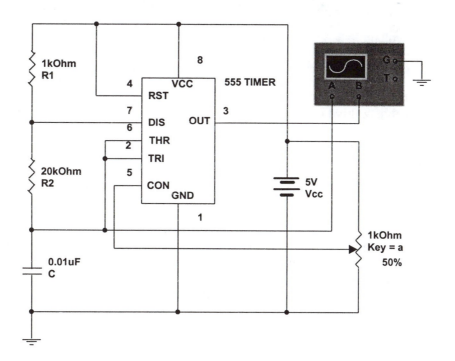

Procedure:

Step 1.   Open circuit file FIG45-1. Bring down the oscilloscope enlargement and make sure that the following settings are selected: Time base (Scale = 200 µs/Div, Xpos = 0, Y/T), Ch A (Scale = 2 V/Div, Ypos = 0, DC), Ch B (Scale = 2 V/Div, Ypos = 0, DC), Trigger (Pos edge, Level = 0 V, Auto). In this experiment you will observe the output of a 555 timer wired as an astable multivibrator (pulse generator). Run the simulation to completion.

Questions: What waveshape did you observe at the output (blue)? Was the voltage across capacitor C (red) that of a charging and discharging capacitor?

Step 2.   Measure the time period (T) for one cycle of the square wave, the output high time (t_H), the output low time (t_L), the upper trip point (UTP) voltage, and the lower trip point (LTP) voltage. Record the answers.

$T =$ _____ $t_H =$ _____

$t_L =$ _____   UTP voltage = _____

LTP voltage = _____

Step 3.   Based on the time period (T) measured in Step 2, determine the output pulse frequency (f).

Step 4.   Based on the output high time (t_H) and the time period (T), determine the duty cycle (D) of the output pulses.

Question: What was the relationship between the output high time and the output low time? How does this relate to the duty cycle?

Step 5.   Based on the circuit component values, calculate the expected output high time (t_H) and the output low time (t_L).

Question: How did the calculated output high time and output low time compare with the values measured in Step 2?

Step 6.   Based on the calculated values for t_H and t_L, calculate the expected output pulse frequency (f).

Question: How did the calculated output pulse frequency compare with the measured frequency determined in Steps 2 and 3?

Step 7.   Based on the calculated values for t_H and T, calculate the duty cycle (D).

Question: How did the calculated duty cycle compare with the measured duty cycle determined in Steps 2 and 4?

Step 8.   Based on the supply voltage $V_{CC,}$ calculate the expected upper trip point (UTP) and lower trip point (LTP).

Question: How did the calculated UTP and LTP compare with the measured values in Step 2?

Step 9.   Change the value of resistor R_1 to 20 kΩ. Run the simulation to completion again. Measure the time period (T) for one cycle of the square wave, the output high time (t_H), the output low time (t_L), the upper trip point (UTP) voltage, and the lower trip point (LTP) voltage. Record the answers.

T = _____ t_H = _____

t_L = _____   UTP voltage = _____

LTP voltage = _____

Questions: Did the UTP and LTP change from the values in Step 2?

Did the value of t_H or t_L change from the values in Step 2? **Explain why.**

Step 10.   Based on the time period (T) measured in Step 9, determine the output pulse frequency (f).

Question: What happened to the pulse frequency when the value of resistor R_1 was changed? **Explain.**

Step 11.   Based on the new value for R_1, calculate the expected value of t_H.

Question: How did the calculated value of t_H compare with the measured value in Step 9?

Step 12.   Calculate the new duty cycle (D).

Question: What happened to the duty cycle when the value of resistor R_1 was changed? **Explain why.**

Step 13.   Change the value of capacitor C to 0.02 μF. Run the simulation to completion again. Change
              the oscilloscope settings as needed. Measure the time period (T) for one cycle of the square
              wave, the output high time (t_H), the output low time (t_L), the upper trip point (UTP) voltage,
              and the lower trip point (LTP) voltage. Record the answers.

              T = _____ t_H = _____

              t_L = _____   UTP voltage = _____

              LTP voltage = _____

Question: What changed from the readings in Step 9? **Explain why.**

Step 14.   Based on the time period (T) measured in Step 13, determine the output pulse frequency (f).

Question: Did the frequency change when C was changed? **Explain why.**

Step 15.   Open circuit file FIG45-2. Make sure that the oscilloscope settings are the same as in Step 1. In this experiment you will observe the output of a 555 timer wired as an astable multivibrator as the control (CON) input voltage varied. Run the simulation to completion.

Step 16.   Measure the time period (T) for one cycle of the square wave and record the answer.

             T = _____

Step 17.   Based on the time period (T) measured in Step 16, determine the output pulse frequency (f).

Step 18.   Click the mouse arrow in the circuit window and type the "A" key on the keyboard to increase the 1 kΩ potentiometer % setting to 80%. This will increase the voltage on the 555 timer control (CON) input. Run the simulation to completion again. Measure the time period (T) for one cycle of the square wave and record the answer.

             T = _____

Step 19.   Based on the time period (T) measured in Step 18, determine the output pulse frequency (f).

Question: What happened to the pulse frequency when the control voltage was increased? **Explain why.**

Notes on Using Electronics Workbench Multisim

1. If you wish to remove a component from a circuit, disconnect both terminals from the circuit; otherwise, you may get an error message.

2. If you wish to pause a simulation on the oscilloscope screen, click the *Pause* symbol (11) next to the switch. To resume the paused simulation, click the *Resume* symbol (11) next to the switch.

3. If you wish to measure the resistance of a component using the multimeter, remove the component from the circuit and connect the multimeter across the component terminals. Make sure you connect a ground symbol to one of the component terminals.

4. The circuit disk provided with this manual is write protected; therefore, you cannot save a changed circuit to the disk. If you wish to save a changed circuit, you must select *Save as* in the File menu and save it on the hard drive or another disk.

5. The color of an oscilloscope curve trace is the same color as the circuit wire connected to the oscilloscope input. A wire color can be changed by right clicking the wire with the mouse and selecting *Color* from the menu on the screen.

6. You can change a component value by double clicking it with the mouse and changing the menu value using the keyboard.

7. You can bring down an instrument enlargement by double clicking the instrument with the mouse.

8. Batteries do not have resistance. To convert an ideal battery (no internal resistance) to a real battery (has internal resistance), add a small resistor in series with the battery.

Bibliography

Boylestad R. and Nashelsky, L. *Electronic Devices and Circuit Theory*. 8th ed. Upper Saddle River, NJ: Prentice Hall, 2002.

Floyd, T. L. *Electronic Devices*. 6th ed. Upper Saddle River, NJ: Prentice Hall, 2002.

Malvino, A. P. *Electronic Principles*. 6th ed. New York: McGraw-Hill, 1999.

Paynter, R. T. *Introductory Electronic Devices and Circuits*. 6th ed. Upper Saddle River, NJ: Prentice Hall, 2003.